CONVICTS, COAL,
and the
BANNER MINE TRAGEDY

CONVICTS, COAL, *and the* BANNER MINE TRAGEDY

ROBERT DAVID WARD
and
WILLIAM WARREN ROGERS

The University of Alabama Press
TUSCALOOSA AND LONDON

Manufactured in the United States of America

Library of Congress Cataloging-in-Publication Data

Ward, Robert David.
Convicts, coal, and the Banner Mine tragedy.

Bibliography: p.
Includes index.
1. Convict labor—Alabama—Birmingham. 2. Coal mines and mining—Alabama—Birmingham—Accidents. I. Rogers, William Warren. II. Title. III. Title: Banner Mine tragedy.
HV8929.A22W37 1987 365'.65 86-4327
ISBN 0-8173-0304-9 (alk. paper)

British Cataloguing-in-Publication
Data is available.

0-8173-1213-7 (pbk: alk. paper)

To the memory of

Jack E. Kendrick,

friend and master teacher.

Contents

Illustrations

Acknowledgments

The authors acknowledge all errors of fact in this monograph as their responsibility and accept the blame for areas that should have been covered but were not. We credit whatever merits this book may have to the many people and institutions who made research material readily available and who often extended aid far beyond professional and functional obligations.

We wish especially to thank the following people for their help: Malcolm MacDonald, Jerrell H. Shofner, Wayne Flynt, Sarah Woolfolk Wiggins, Douglas Helms, Bawa S. Singh, Richard Wood, John Hebron Moore, Betsy Frank, Eugenia Rankin, Miriam Jones, the late Milo B. Howard, Jr., Edwin C. Bridges, Mary Louise Ellis, Joseph Evans, Kathryn E. Holland Braund, William Warren Rogers, Jr., Phyllis Holzenberg, Marianne Donnell, Joann D. Boyer, Gayle Peters, Robert Overstreet, Marvin Whiting, John W. Hevener, Diane Jackson, Jane E. Keeton, Floyd Watson, and David Ruffin.

We appreciate the cooperation and aid furnished by the staffs of libraries, archives, and depositories of public records at Georgia Southern College; Florida State University; Auburn University; University of Alabama; University of North Carolina; the National Archives, Washington, D.C., and the branches of the National Archives at Suitland, Maryland, and East Point, Georgia; the United States Department of Interior Library, Washington, D.C.; Birmingham Public Library; Carnegie Public Library, Eufaula, Alabama; Georgia Department of Archives and History, Atlanta; Alabama Department of Archives and History, Montgomery; Labor Archives, Georgia State University; Jefferson County Courthouse Files, Birmingham; Library of the State Supreme Court of Alabama, Montgomery; and Bureau of Vital Statistics of Alabama, Montgomery.

The authors are grateful to Jane Ward and Miriam Rogers for being patient and impatient at the right times.

ROBERT DAVID WARD
WILLIAM WARREN ROGERS

CONVICTS, COAL,
and the
BANNER MINE TRAGEDY

Introduction

The story of the Banner Mine disaster on April 8, 1911, and the events that followed cannot be understood without an explanation of earlier conditions and developments. A complex mix of politics and economics was involved. Reconstruction ended in Alabama with the political triumph of the Bourbon Democrats in the gubernatorial election of 1874. The old planter class of Democrats and former Whigs installed themselves at the state capital in Montgomery and put into effect their philosophy of white supremacy and of economy and efficiency in government. Shortly afterward, a new set of powerful men began to rise in such emerging cities as Birmingham, Anniston, Gadsden, and Bessemer. Their political and economic conservatism melded well with that of the leaders from the agricultural regions. Despite rivalries and occasional differences, the two groups struck a mutual alliance and ran the state, acting out in classic form the credo of the New South.

The Bourbon Democrats controlled Alabama politically. Yet economic power was shared, especially in the industrial cities and towns of the north-central mineral belt, by both native sons and Northerners. Many of the latter were Republicans. These men were far from being ostracized, for they represented Yankee capital and industrial know-how, welcome commodities in an impoverished agricultural state.

In the rest of Alabama little had changed. During the last half of the nineteenth century and well into the twentieth, a majority of the people earned their living from the land. The state's yeoman farmers (among whom there were a limited number of blacks) struggled to survive by the farming of one crop—cotton. They did so in the face of rising surpluses, falling prices, and increasing costs of production. The most destitute of all were the sharecroppers, mostly black, although they were joined by more and more whites. The tenant farmers worked land and paid their rent to the landlords in money received from the fall sales of cotton. Supplies were obtained from commission merchants, and they also were entitled to a portion of the crop money. The destiny of the sharecropper and his family was determined by a small piece of paper known as a lien note. The once rich but now depleted soil received its spring ritual: a boost of expensive fertilizer. But dreams of profit were replaced by hopes of breaking even, which gave way to the reality of deeper debt with every passing year. By 1900, Alabama's rural population had little hope.

By way of contrast, in the late 1870s Jefferson County and the town of Elyton (the future Birmingham) became the focus of a remarkable industrial and mining revolution. Together with the surrounding mineral-rich counties, the area was penetrated by railroads. Surprisingly large deposits of bituminous coal, limestone, and iron ore—the exact ingredients for the manufacture of iron and, later, steel—began to be exploited. Now, with transportation, modern extractive techniques, and capital, the region's geological riches—endlessly described as being "locked in the bowels of the earth"—began yielding enormous profits.

The entrepreneurs who gravitated to the region were colorful, competitive, and highly individualistic. They formed and reformed companies and corporations, befriended and betrayed one another, and amassed fortunes in iron and steel manufacturing and in coal mining. Incongruously, part of Alabama became industrialized but remained juxtaposed on the north and south by larger rural areas. A perceptive northern scholar observed as late as 1934 that "Birmingham is not like the rest of the state. It is an industrial monster sprung up in the midst of a slow-moving pastoral. It does not belong—and yet is one of the many proofs that Alabama is an amazing country,

Introduction

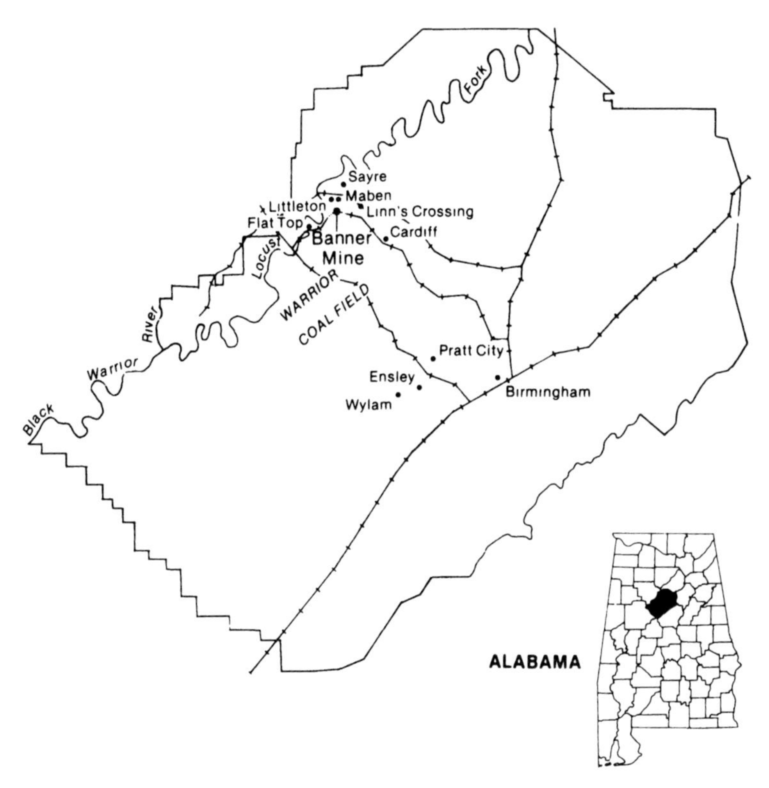

Jefferson County in 1911. (Map prepared by Beverly Minor)

heterogeneous, grotesque, full of incredible contrast. Birmingham is a new city in an old land."[1]

The state remained under the political dominance of the Bourbon Democrats, although there were scattered but ineffective Greenbacker and Independent movements in the 1870s and 1880s. Far more serious was the challenge of the Populists in the 1890s. The People's party was the political outgrowth of economic discontent by the state's farmers. The first evidence was the Farmers' Alliance, organized from the local to the state and regional levels (the southern and northern alliances were separate). In Alabama and the South there were white and black alliances and some that contained both races—finally some desperate agrarians realized that a crop lien note was color blind. When its ambitious social, educational, and economic programs failed, the Farmers' Alliance moved into political action. The result was the bitter fratricidal warfare of the 1890s. Before the Populists were defeated by appeals to racism, political treachery, and their own lack of solidarity, they sent tremors through the power bases of the Democrats. Alabamians were amazed at the Populist combination of whites and blacks, farmers and coal miners (representing organized labor), Republicans and ideological dreamers—all fed up with Bourbon chicanery.

After 1896 the Populists faded from the picture only to be replaced by the Progressives. Advocating regulation rather than profound change, the Progressives tended to be middle class and avoided exposing themselves in the trenches of radical reform. Still, they managed to make an impact that, all things considered, was more positive than negative.

In a broad sense, this book is concerned with all the developments mentioned in this introduction. Yet its emphasis is on the flawed implementation of the New South as a concept. Industrialization, New South proponents insisted, would produce an economic utopia. Yet only a few persons shared in the resulting benefits. More specifically, the book is about the men who worked involuntarily in the Banner Coal Mine, owned by the Pratt Consolidated Coal Company. And it is about the repercussions and consequences that followed from an explosion at the mine in the spring of 1911.

A labor force was necessary to maintain and expand the Birmingham area's industrial boom. Many of the workers were native Alabamians—blacks and whites who moved from the debt-ridden agricultural regions to the steady paying jobs in the coal mines and blast furnaces. There was as well an ethnic work force, small but important, from England, Wales, Germany, Italy, Ireland, and elsewhere. The men and their families were both second-generation settlers. As one historian has written, such people were professionals who took pride in their work. For them "the shared dangers, the craft identity, the rhythm of mine operations promoted an abiding solidarity that was deeply rooted in an identification of the past."[2]

The native and immigrant free labor force was not an unmixed blessing, especially when the workers began affiliating with labor unions and striking for higher and better working conditions. Yet the owners of the coal mines also had available a reliable but completely artificial work force: convict labor, both county and state. Alabama's state and county officials sought to avoid expense and to reap financial profit by leasing their prisoners to coal mines, lumber camps, and farms. In the post-Reconstruction decades Alabama's prison population soared. The prisoners, most of them blacks, represented an expensive problem of maintenance. A pragmatic solution to the difficulty came in the form of the convict lease system. Businessmen hired the convicts from the state and various counties, paying a fixed rate for their labor.

The Banner Mine used convict labor and was a visible expression of Alabama's penal system that leased its prisoners to private business. This policy was a major social issue in its own right, the scene of countless battles as change and reform clashed with a deeply entrenched institution. As a result, the awful accident at the Banner Mine presented the public and its representatives with issues that confused reaction and that forced a departure from any predictable response.

No attempt is made in the narrative that follows to provide an in-depth study of Alabama's convict lease system. There is no detailed account of coal mining, corporate practices, or political machinations—although much of all these elements is included. The full story of the convicts themselves cannot be told because biographical information on them is almost nonexistent. The explosion that shattered the work routine at the Banner claimed 128 lives, mostly those of black convicts. Tucked away in the rugged hills of northwest Jefferson County, a few miles from Birmingham, the Banner was a highly productive mine in the rich Warrior coalfield (named for the river that drained its basin). The tragedy that occurred there was important. It deserves to be viewed and remembered as something more than a small and fleeting event on a vast and teeming stage.

1
Death at the Banner

It was Saturday morning, April 8, 1911, another day to be marked off by the convicts in the Banner Mine prison. Days and weeks and months were the slow but cumulative units of progress that eventually could bring freedom from the coal mine.

That spring morning in the hills of north Alabama the sun first rose at 5:39. It was wrapped in clouds and mist. Those leaving the mine from the night shift would have cool sleeping, and deep in the mine, the rain would not disrupt the incessant labor of the day gang.[1]

Before the sun came up the guards had aroused the convicts on the day shift. They left their peculiar swinging beds and ate their breakfast; they made their preparations for the last shift of the week before Sunday finally brought a day of rest. At 5:45 they marched out of the prison enclosure and entered the 1,700-foot chute that ushered them like cattle from prison wall to mine shaft entrance. They walked quietly, inured to another ten hours of back-breaking labor. The burden of another day's quota of coal would be borne, as usual, with stoic acceptance.[2]

At 6:00 the night shift left the mine. Night fire boss William Sparks announced that the mine was in good condition. Mine boss John Cantley was not present; thirty minutes later he still had not entered the shaft. As the convicts filed past, the free laborers who ran the cutting machines lounged around the entrance. The shot firers collected their supplies of bituminite and fuse and paper and prepared to follow the convicts into the mine.[3]

Convict foreman O.W. Spradling, a veteran who had worked with

A cluster of buildings surrounded by a wooden wall composed the prison at the Banner Mine. (Birmingham Public Library Archives)

prisoners for twenty years, led his day shift into the shaft and issued his orders. The convict miners dropped off in the side galleries to start their work. They went into four left, five left, six left, seven left, and still deeper into the mine that already ran a mile into the earth. There were five free men in the mine that morning. Foreman Spradling, fifty years of age, lived with his large family in the town of Leeds. The other free men were the shot firers. They handled and fired the explosive bituminite that blasted the coal from the face, permitting the fragments to be loaded into the cars. The shooters were Lee Jones, white; Mose Lockett, black; Dave Wing, black; and Daddy Denson, black. They went down the shaft, and they may have stopped near seven left. Near the shooters was John Wright, a white convict, doing electrical work in the mine as legal penance for assault and battery on a female.[4]

The moment had arrived. At 6:30 an explosion and blast of flame occurred near seven left. John Wright was blown to pieces, and the four shooters were killed instantly. The next moments were that surreal interim of emergency when, for participants, time slowed down and every motion was magnified. There was noise, and the air was filled with dust. The miners near seven left entry heard and felt the explosion, while those deeper in the mine had no warning of danger. The huge twenty-foot Crawford-McCrimmon fan, recently installed at a cost of three thousand dollars, was blown out by the explosion. The steady movement of fresh air through the mine slowed and stopped, and the auxiliary fan, far away in number one shaft, did not come on.

Clarence Nicholson, working in five left, knew there had been an explosion. Experience told him that time was critical and that the deadly blackdamp would soon flow through the corridors, killing everyone in its path. The black convict, who could have run to safety, raised the alarm as he moved deeper into the mine, yelling a perhaps less stilted version of the reported "Get out, men, or you will all be destroyed." Eight men fell in behind Nicholson, and others joined them. They raced for the shaft, and probably forty made it unharmed.

Another convict, Charley Brown, heard the yells and ran to safety but went back to lead twelve other miners out. Warned by Nicholson, Curlie Smith escaped. Like his fellow convicts Nicholson and Brown, Smith reentered the shaft and guided three miners to safety. James Franklin, a black convict serving time for grand larceny, heard Nicholson's cries, dropped his pick, and started running. He saw Nicholson collapse, and he had a "horrible feeling . . . a most horrible feeling." Franklin found his way blocked by fallen coal and rock and turned back. Nicholson somehow revived and, according to Franklin,

> He urged me to try to get through some old work and I grabbed up my pick and started to work with him. The blackdamp came on us. I felt it. We renewed our efforts; we pushed on hard. Nicholson was about to give up when we succeeded in getting into a new air course and then we struggled out. It was an awful experience, believe me, an awful experience. I didn't believe I was going to get out.[5]

As Nicholson and Franklin struggled to escape, J. T. Massengale, the white assistant foreman, was able to lead sixteen men out.[6]

J. Flavious Erwin, agent for the Louisville and Nashville Railroad (L & N), maintained communications for the line between the mine and Birmingham, twenty-five miles to the east. He left his home that morning to walk to work. It was raining lightly as he strolled along, but there was little cause to hurry. There were records to check and paperwork to complete, but the morning train from Birmingham was still hours away. As Erwin neared the Banner he heard the rumble of an explosion. In confirmation of his hearing, Erwin saw the ventilating fan stop, its constant roar now extinguished. He first ran to the mine's entrance to see if he could help; then Erwin hurried on to his office to send the first news of the explosion to the outside world. The agent tapped out his message to W. F. Wyre, agent in Birmingham: "Banner Mines were blown up just now, about two hundred men in there and killed." If Erwin was in error in detail, he was quite correct in scope and magnitude.[7]

Fifteen minutes after Wyre received the message from Erwin, the news was delivered to the Brown-Marx Building, which housed the corporate offices of the Pratt Consolidated Coal Company. Although the company learned of the explosion as early as 7:00 and no later than 7:30, it did not notify the office of the state mine inspector nor that of the United States Bureau of Mines until 8:30. On notification, Angus B. Brown, in charge of the Bureau of Mines office in Birmingham, immediately sent a telegram to Chattanooga, Tennessee, requesting the dispatch of rescue car number six. The car, specially equipped by the Bureau of Mines, carried respirators that would allow rescuers to penetrate the gas and dust in their search for survivors. In any event, it took 16½ hours for the car to travel the 150 miles from Chattanooga. The train stopped at Mineral Springs to load mine timber and "the needed equipment for building brattices," which probably accounted for what otherwise was described as "a record run from Chattanooga."[8]

If Bureau of Mines aid was far away in miles and time, there was a closer source of mine rescue equipment. Even before Brown wired for the Bureau of Mines car, the officials of the Tennessee Coal, Iron, and Railroad Company (TCI) heard of the explosion and offered to send their rescue car and trained men to the scene. Their offer was declined by the Pratt Consolidated—the official who made the decision was never named—on the grounds that it would be to no avail. The conclusion was correct, but it was drawn before there was sustaining evidence.

Not until 9:30 that morning did an L & N train leave Birmingham bound for the stricken Banner. Aboard were James Hillhouse, the state mine inspector; his son James Hillhouse, Jr.; and T. W. Dickinson, an assistant state mine inspector. Twenty years earlier the avuncular Hillhouse had come into the Birmingham area and had directed coal mining operations for different companies. "Uncle Jimmy" was a dedicated mine inspector and, of necessity, had wide experience in coping with mine disasters. The fifty-seven-year-old Dickinson had twenty-three years' experience with mines and had dug coal himself. He had the respect and confidence of the miners.[9] Also on board were officers of Pratt Consolidated: E. P. Rosamond, general superintendent of mines, and H. E. McCormack, vice president and general manager. Neither Erskine Ramsay, another vice president and general manager, nor George B. McCormack, company president and brother of H. E., was present.

The explosion that apparently had claimed the lives of an undetermined number of miners, most of them black convicts leased from counties across the state, would inevitably have far-reaching ramifications. The accident's occurring at the Banner was particularly disquieting. The Banner, as its name implied, was the safest mine in Alabama. It was owned and managed by men of impeccable credentials. Personnel in the Brown-Marx Building were shocked. Could it really be true? Pratt Consolidated was Alabama's largest miner and seller of coal (its holdings included fifty-four mines and eighty-five thousand acres of land in Alabama and Tennessee).

A consolidation of six companies, the corporation was formed in 1904. At fifty-two, George B. McCormack was at the height of his energy and influence. He was a Scotch-Irish native of Missouri with a flair for inventiveness and a capacity for hard work. Intense and sharp-featured, he had moved to Tennessee and had begun working for the Tennessee Company, which became TCI, and then was assigned to Birmingham. McCormack had layers of thick straight hair, an aquiline nose, and a mustache and was a formidable executive, a master of detail. He rose to top management posts with TCI before forming the Pratt Coal Company in 1896 and Pratt Consolidated in 1904. By 1911, he was also president of the Alabama Coal Operators Association and was renowned for his common sense. A contemporary wrote that he had "those qualities of cool judgment, practical wisdom, and personal force" that made him "the great captain of modern industrial Alabama."[10]

With his quiet demeanor, low-key air of confidence, conservative dress, and balding head, Erskine Ramsay might well have been mistaken for a professor of classics. Instead, the forty-seven-year-old Ramsay was a gifted mining engineer. Of undiluted Scottish descent, Ramsay was Pennsylvania born and college educated. His father was the chief mining engineer for the powerful Pittsburgh industrialist Henry Clay Frick. Young Ramsay came to Birmingham in 1887 and became an associate and friend of George McCormack. Ramsay also ascended to the upper echelons of power within TCI and patented several significant inventions. After fifteen years with TCI he joined Pratt Consolidated.[11]

H. E. McCormack was two years younger than George. The journey from Missouri that brought him to Birmingham was similar to that of his brother, although he had once searched for gold in the Dakotas. Less aggressive than George, the younger McCormack was described as "a most indefatigable worker, keen, quiet, and alert."[12]

Ramsay was the Pratt Consolidated official most disturbed by the news from the Banner. The mine had opened in 1904 and was the special interest of the brilliant mining engineer. He had carefully set up the mine, and within the company it was known as his "pet." The Banner was the first mine in Alabama to install electric lighting, electric haulage, and electric coal cutting.[13] And yet the unthinkable

George B. McCormack, powerful president of Pratt Consolidated and spokesman for coal mine owners. (Alabama Department of Archives and History)

Erskine Ramsay, vice president and chief engineer of Pratt Consolidated. He made the Banner Mine a showplace of modern mining techniques. (Alabama Department of Archives and History)

had happened. The officers would have to deal with the situation. They had dealt with other crises, but for the moment they did not even know the facts.

As the L & N train left Birmingham to climb the hills to the Banner, events at the mine moved from immediate shock to response and positive effort. L. W. Friedman, a reporter for the *Birmingham News,* gained a seat on the train, and his stories would be the first accurate accounts of the disaster.

Most of the forty men who managed to escape reached the main shaft and from there were led to safety by convict Charley Brown and Assistant Foreman J. T. Massengale. But following that first evacuation, only Clarence Nicholson and James Franklin appeared at the shaft. Now the mine was quiet, with only shifting dust in the air to mark the recent explosion. Somewhere in the Banner more than one hundred men still lived if they had avoided the blackdamp, or they were dying from the gas, or they were already dead. Outside the mine there were only questions and conjectures. The answers lay inside along the corridors of the Banner.

Two men made the initial effort at exploration. Clark McCormack, the son of H. E., accompanied by former miner J. R. Baird, went into the mine. Their first discovery was less than reassuring. Not far from the mine entrance and pathetically near to safety, they found a man, seated but leaned forward and quiet as though asleep. He was convict boss O. W. Spradling, dead from the blackdamp. The air was foul, but McCormack and Baird pushed on. Next, they found "a rough pile" of dead men and dead mules, all apparently killed by the blast.[14] The two rescuers, now choking and dizzy in the suffocating air, could stand it no longer. They turned back and staggered to the surface; it was lucky that two more fatalities had not been added to the toll.

When the rescue train arrived from Birmingham, a council of war was held between the company officials and Hillhouse and his rescue team. Clark McCormack and Baird reported that falls of rock blocked some of the corridors, and the high concentrations of gas were duly noted. There was scant reason for optimism. But if reason argued that all were dead, it was still possible that somewhere in the twisted tunnels some men still lived and waited, crouched and fearful and clutching at the hope of rescue.

Inspector Hillhouse and his six rescue men had changed into overalls and jackets on the train and were ready to penetrate the Banner. They were accompanied by sixteen convicts who volunteered for the mission. The twenty-three men entered the mine and in single file moved quietly down the shaft. They came to the body of Spradling, paused, and then moved on. Then they reached the pile of men and mules reported by McCormack and Baird, and they went no farther. The air was suffocating. With heaving chests and aching heads they loaded the bodies on a tram car and headed for the surface. Surrounded by only a small crowd, silent and subdued, the Hillhouse team unloaded the bodies of Ernest Knight, William Garth, and Columbus Nave—the first of a four-day procession of bodies that left the Banner.[15] Sam Lively, one of the convicts who had accompanied Hillhouse, collapsed at the surface and was taken away to the prison hospital.

With little rest or recuperation the Hillhouse team reentered the mine. The gas remained, but the men pushed even deeper, and by 1:00 on Saturday afternoon they had returned to the surface with three more bodies. By this time news of the disaster was spreading, and a crowd of several hundred from neighboring mining communities clustered around the mine shaft. But the tableau common to countless other mining disasters was missing at the Banner. These were convict miners, and their families were far away. The crowds, more detached than bereaved, stood in the rain and watched with professional interest as the rescue efforts continued.

As Saturday afternoon wore on, those rescue efforts began to slacken. The high gas concentrations left the rescuers increasingly groggy, and Inspector Hillhouse had to be brought to the surface twice after lapsing into unconsciousness. It seems now far better strategy to have concentrated efforts on repairing the ventilation fans and building brattices to direct the airflow so that the rescuers could enter the mine with some safety. Although it was not publicly stated, it was clear to the decision makers that there was no chance for any of the miners to have survived. Mine repair and body removal were the actual jobs at hand. But as the sun began to set on Saturday there was speculation among those who watched and waited that a few men might still be alive. President McCormack made his judgment public with a pronouncement that, if any of the miners were still alive, "it would be by intervention of some power more than human."[16] That at least narrowed the possibilities.

At 1:35 A.M. on Sunday morning the Bureau of Mines's rescue car finally arrived from Chattanooga. Dr. J. J. Rutledge was in charge of the car, and he was joined by Angus Brown of the Birmingham bureau. Rutledge and Brown immediately placed themselves at Hillhouse's service, and a conference on strategy ensued. But if the train brought this additional professional help, it also brought the image makers and the politicians. Three members of the State Convict Board, headed by President James G. Oakley, were also in attendance. Oakley had been at Centreville (the county seat of Bibb, a peripheral mining region) when he received a telegram from Governor Emmet O'Neal. The terse communication informed him that there was an "explosion at Banner Mines today killing over one hundred" and ordered him to "go at once and investigate."[17]

Oakley, who lived in the small town of Ashby in Bibb County, had been on the job only a little over a month. He was serving as sheriff of Bibb County, when on March 4, O'Neal had appointed him president of the Board of Convict Inspectors. The governor had known his appointee over the years, although "my acquaintance with Mr. Oakley had not been intimate." The appointment came because the candidate mounted a well-organized campaign of support. As O'Neal said of Oakley, "Almost daily I was besieged by prominent citizens of the State, who personally called to urge his appointment."[18] O'Neal could not know it, but choosing Oakley would prove harmful to the state and to his administration.

When Oakley arrived at the Banner he surprised officials and onlookers by donning overalls and going into the mine for an inspection. Emerging shortly, Oakley generalized that an explosion had interrupted the airflow, causing wholesale suffocation. Associate Inspector Hugh M. Wilson, a well-known politician and influential editor, reacted with personal horror, but his thoughts also turned to the state capital. Had the accident occurred two weeks earlier when the legislature was debating a new mine safety law, Wilson declared, there would have been a clamor to remove the convicts from the mines of Alabama. That threat, by implication, was now behind them.

The members of the convict board, an appearance made, soon left the Banner. The focus returned to the work of rescue. The presence of both state and federal officials raised the possibility of jurisdictional problems, but the geography of the Banner allowed a simple solution. Hillhouse and his state rescue team would continue to work from shaft number one; Rutledge and his federal men would operate from the new shaft. With almost a mile between them, there were no problems of precedence and procedure.

At 3:00 A.M., with Oakley and his colleagues retired from the scene, Dr. Rutledge made his first assault on the Banner. The auxiliary fan was now running, and Rutledge, accompanied by three men from the Bureau of Mines and five other volunteers, began his descent. Slowly and cautiously the men worked their way along the corridors. As the group approached the fourth left entry, they began to stumble and fall. A boiler accident had stopped the steam-powered fan, and gas concentrations had immediately built up. As Dr. H. H. Hamilton recalled, "We were going along all right until we got to the fourth entry, and in a few seconds nearly every man in the party was down. I attempted to carry one man over the lift, but failed and fell myself. That is the last I remember until the surface was reached." Only Dr. C. N. Carraway was still on his feet. The official turned, dragged himself back to the shaft entrance, and cried, "Get us out; we are dying." Another team, headed by Dr. Wright of the Bureau of Mines, immediately put on "oxygen helmets" and hurried to the rescue. Back on the surface it first seemed that Dr. Rutledge

A shaft at the Banner Mine. (Birmingham Public Library Archives)

and two others were dead, and "physicians had to work over them for nearly one hour before they were able to be moved."[19] The Banner had almost claimed an additional dozen men.

The close call of the Rutledge group had a chastening effect on all rescue efforts. Once again the logical conclusion had to be reached: there was no longer any hope that anyone remained alive in the mine. All rescuers would be withdrawn and efforts would be concentrated on establishing fan ventilation to clear the gas. The rock falls could be removed and the air courses reestablished. Inspector Hillhouse made the decision to reverse the ventilation fan in order to push fresh air into the mine, in preference to its usual exhaust function. The change in air direction provided a greater airflow, and as the fan roared away, all rescue efforts halted.

It was inevitable, and not long delayed, that speculation on the causes of the explosion would flourish. The early evidence from the rescuers showed that the bodies near seven left had been badly mangled—"Parts of one white body were found scattered for several yards"—while those farther away were undamaged. In addition, it was reported that the tracks at seven left were severely twisted and that nine tram cars had been "hurled to the end of the track and broken into fragments."[20] Pending further evidence, there was agreement that seven left had been the scene of the explosion. There was much less agreement on the cause.

All of the possible causes of mine explosions were rehearsed: the use of blasting powder had ignited a coal dust explosion; a "sudden feeder" of gas had entered the mine and had exploded; the blasting powder had been handled near an open lamp.[21] It was recalled that in 1910 three men had been killed at the Banner when their blasting caps and bituminite had exploded. Had it happened again on a larger scale?

The most fully developed story of causation held that John Wright, the white convict concerned with electrical repair, had accidentally sparked two wires together and thus set off the explosion—whether of gas, dust, or bituminite was not specified. Such evidence was both imaginative and circumstantial, but it gained some credence from the report that Wright's body and those of the shot firers were found in close proximity.[22]

There was yet another possibility, hinted at but not explored in the public press. Since its opening the Banner had been listed as a gassy mine. If there had been any interruption in the ventilation system, it might have allowed dangerous levels of gas to build up. It was exactly this line of speculation that the Pratt Consolidated moved to block. The question of responsibility for the disaster opened the issue of claims made by the families of the dead convicts. The Pratt Consolidated immediately attempted to clear itself of any hints of negligence.

After the explosion, fan tender Browning, in charge of the huge fan at number two shaft, got the records for the previous night and held them until he could talk with Mine Superintendent J. S. Waldrop. According to a reporter, those records told "a tale of steadiness" on the air supplied during Friday night, although it was not clear that the reporter actually saw the records.[23]

The initial statement by company officials came from Chief Engineer Erskine Ramsay. He was tentative in his conclusions, as befitted the paucity of evidence. Ramsay's remarks were simply his reactions, not an official statement from the company. The seventh left entry seemed a likely place for the explosion, but Ramsay was "unable to say" if the explosion had been caused "by powder or damp." It did seem likely that most of the miners had died of afterdamp, and it was true that mine damage totaled no more than $1,200.[24]

Two days after the explosion, and long before the mine inspectors had made their investigation and other official reports were released, George B. McCormack made a lengthy formal statement on the accident. The president of the Pratt Consolidated emphasized the positive. The mine was run with "every precaution which modern and improved mine appliances could suggest." The fans were in perfect condition and were running at the time of the explosion. There were two "experienced and competent fire bosses." The night fire boss had "made a thorough inspection of the mine" and had just left at the time of the explosion. Since the mine was dry, it was sprinkled daily to prevent the accumulation of coal dust, and it was in fact "wet and muddy" after the accident. It could not have been a dust explosion, said McCormack, and with the ventilation provided, "It is incredible to us that the explosion was caused by gas." McCormack argued that only a little gas "has ever been generated in the mine." If, in spite of all indications to the contrary, the explosion had been caused by gas or dust, "The company officers were absolutely unable to foresee it, and powerless to avoid it."

So what indeed had caused the explosion? The company used only permissible explosives and absolutely forbade a convict to fire a shot. In fact, "All of the shot-firers were free white men." So, said President McCormack, "The only way I can account for the explosion is that someone did something about the powder that caused it to explode."[25] The negligence was individual; it most definitely was not corporate.

In his understandable desire to exonerate the Pratt Consolidated, McCormack twisted his evidence at several points and on occasion misstated it. It was not true that the explosion took place "between the time the night crew came out and the day crew went in." If that had been the case, the only fatalities might have been the few supervisory personnel who were still in the mine. It was plain that the day shift had been in the mine a full half hour before the explosion occurred. Nor was McCormack accurate in his remark that "all of the shot firers were free white men." As previously identified, the four shot-firers were free men, but three of them were black, a fact McCormack may have thought prejudicial to the safe handling of explosives.

While McCormack insisted that "very little gas has ever been generated" in the mine, a spokesman for labor disagreed. "Mr. McCormack," he wrote, "don't you know it is not safe to have electricity in gaseous mines?" It was a matter of record that state mine inspectors had repeatedly listed the Banner as a gassy mine.[26] McCormack's statement was an extended one, but it left many matters in doubt and many not even mentioned. Yet his explanation established the position that the company was to hold throughout the proceedings. Other investigations had yet to be made.

By Sunday morning the rain had ceased, and the sun shone brightly. Now the scene around the Banner Mine became one of bustle and excitement as crowds of the curious began to arrive. These "excursionists" came on the trains of the L & N and the Southern; they came by automobile, jarring along the dirt roads; the less affluent and more practical arrived by wagon and "on horseback and mules."[27] One observer was repelled by the scene and reported that the "sights outside were enough to turn a heart of stone. The day being Sunday, the crowds flocked to Banner by the hundreds, and a holiday spirit seemed to possess the most of them." There, "clustered about on the hills and under the trees were the crowds of sight-seers. The air was filled with jokes and ribald jests, while the air also was heavy with the odor of mean whiskey."[28]

Deputy Sheriff Dave Kennybrook and his guards strung up rope and patrolled their lines with rifles and shotguns in an overzealous effort to enforce order and to keep a way clear for the rescuers. The guards dispersed neighboring free blacks who were searching through the castaway clothes of the dead—they were unaware that the prisoners had no valuables. Through it all photographer Bert Covell of Birmingham took pictures of the crowd and of the yawning mine mouth—its darkness a testimony to mystery and fear and tragedy.

Inspector Hillhouse's strategy of reversing the fans to clear the mine gradually proved successful, and the rescue teams were able to penetrate farther into the galleries. Now tram car loads of bodies began to arrive at the surface, and a rush order was made to the Green Coffin Company of Birmingham for one hundred coffins. Other coffins and shrouds came from the John's Undertaking Company in Birmingham, but the supply was soon exhausted, and a special shipment was obtained from Nashville, Tennessee. After being cleaned in the prison washhouse, the bodies were embalmed by Echols and Angevin of Ensley. Their task was so great that the morticians had to be assisted by some of the surviving convicts. Unless families or next of kin called for the bodies, they would be buried on the mine's premises. A gang of twenty convicts began the process of digging "a long trench in the convict cemetery."[29]

If there was curiosity and the compelling ambience of tragedy, there were few signs of bereavement and loss. One elderly black woman cried for her dead nephew. Another sorrowing black woman wept for her husband in the prison morgue. (There was mistaken

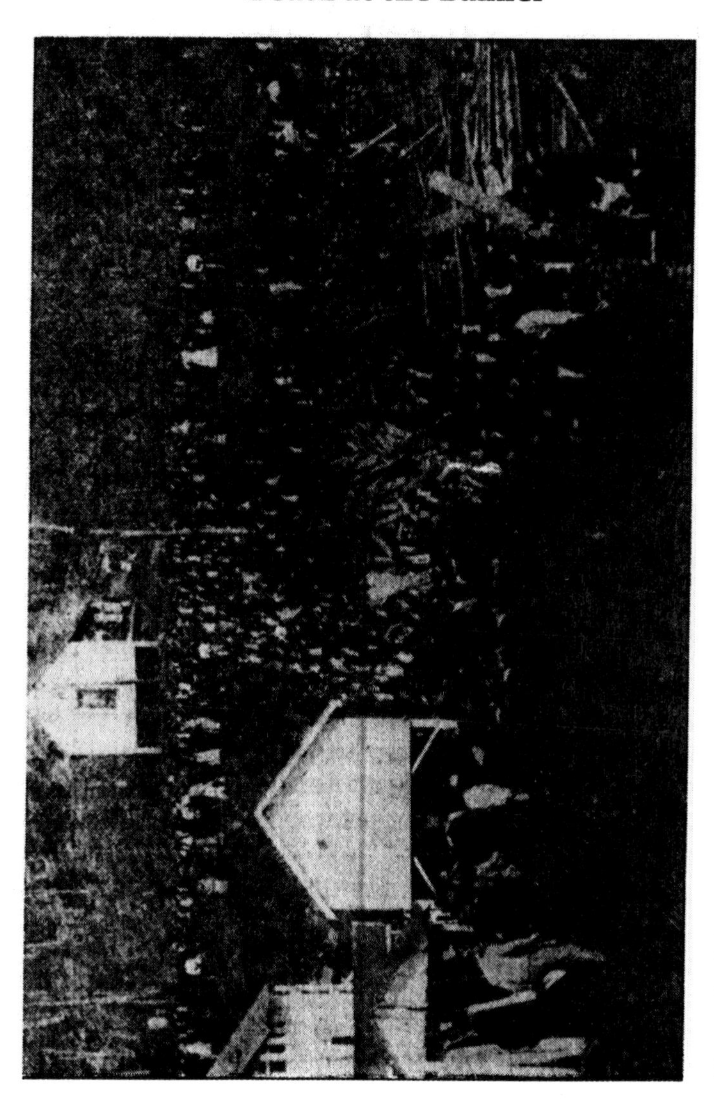

Aftermath of the explosion, as pictured first in the *Birmingham News*, April 10, 1911. Photograph by Bert C. Covell. The headline above the picture said: "128 ARE DEAD AT BANNER: 45 BODIES ARE RECOVERED." (Alabama Department of Archives and History)

identity here, and she was led away still crying.) People expected friends and relatives to appear on the scene, and the ambulance chasers of the legal profession were on hand to suggest a claim against the company. But there was little business to be done at the Banner. (Also, Pratt Consolidated's own lawyer, R. B. Watts, was there to safeguard the company from the dangers of litigation.) The locale contrasted sharply with the event. As the sun sank behind the Alabama hills, soft green with the beauty of spring, the observer who had been critical earlier was reinforced in his cynicism. "The crowds began to divide," he wrote. "Food was scarce, and so was whiskey, and the crowds left with the feeling of having had a good time. The blind-tigers [illegal liquor stands] did a land office business and so did the Pratt commissary, where a thin slice of ham and crackers cost ten cents."[30]

Disaster had publicized the Banner Mine and had brought it a fleeting notoriety. Its list of visitors in a few days far surpassed the number who had ever seen or heard of it. On Monday morning the last of the important and the influential made a courtesy call. Dr. Joseph A. Holmes, the head of the United States Bureau of Mines, arrived in Birmingham from Pennsylvania to make his own investigation and to show the bureau's flag. Accompanied by Erskine Ramsay and Robert A. Brown of the Pratt Consolidated, Holmes, as one newspaper reported, "went through the country in an automobile" to reach the Banner Mine.[31]

Holmes accompanied a crew into the mine (some reports said that he spent seven hours in it) and emerged with a report that it had suffered only slight damage. The Banner, he said, had a modern and efficient ventilation system. Holmes avoided even the appearance of controversial comments, and he made no mention of the use of convict labor. Even on his return to Washington, Holmes confined his remarks to the technical aspects of the disaster and the need for mines to be equipped with fire-fighting apparatus.[32]

Six days after the explosion, the last of the bodies was removed. The total of deaths, fluctuating up and down in the daily newspapers, was finally and officially stabilized at 128. The difficulties of counting had been compounded when so many parts of bodies had been recovered. Two whites and three blacks made up the five free men killed. There were ten whites among the 123 convicts who perished. Prisoners killed at the Banner Mine came from fifteen counties, although Jefferson with seventy-one was far ahead. Other counties that had leased prisoners to Pratt Consolidated included Barbour (two), Butler (five), Calhoun (two), Chilton (one), Escambia (six), Franklin (one), Greene (eight), Henry (one), Marion (three), Morgan (six), Perry (four), Tallapoosa (two), Tuscaloosa (eight), and Walker (four).[33]

A reporter covering the story wrote, "When I viewed the cold faces of the dead black men, I wondered then were it now better for them, for it was certainly the last escape—the one way of beating justice, fate, or perhaps injustice."[34] With the dead removed, it was announced that the Banner would be able to resume operation within ten days. This depended, however, not simply on making the necessary mine repairs but on the question of whether the counties would send in more convicts to resupply the labor force. That question would be answered according to the resolution of the conflicting pressures of material gain and the shock of public opinion.

2
The Convict Lease System

The simple question is: Why were convicts working in the Banner Mine or in any coal mine? The answer is not simple. Every society has confronted the problem of dealing with individuals who will not conform with its rules. Some people create turmoil and strife by personality alone. Some steal and pilfer, and some introduce the cancer of internal violence that no state can long allow. Man is a fractious animal; in their unsocialized behavior, people are a danger and a menace to their fellows.

We have few alternatives in dealing with the criminal. We can separate him from the society that he harms by banishment to another land or by the artificial banishment to an imaginary other land of prisons. We can whip and beat and brand the criminal and use pain as punishment, or despairing any hope and in deep revulsion at his crimes, we can take his life in forfeit. Over time, the answer of what to do has moved through different stages. The stocks of pain and public ridicule were slowly opened as punishment by whip or branding iron became offensive to changing moral sensibilities. Slowly, the list of capital offenses was reduced. In place of pain or execution, the prison became the major answer for either punishment or rehabilitation.

Practical Alabamians began their experiment with state prisons on the firm premise that prisoners must earn their keep. Those Northern thoughts of convicts sitting in their cells and meditating on the enormity of their crimes and the need for redemption were alien and far too expensive for a poor state to entertain. Having met the basic costs of conscience with a prison, it seemed an insult to every working man to support a criminal in sloth and idleness. No one reads the future with precision, and surely no one in the Alabama legislature on January 26, 1839, could have guessed where his enlightened vote to establish a prison would lead.

When Alabama became a state in 1819, it installed the death penalty for an imposing list of crimes from treason to counterfeiting. It treated lesser infractions of the law with fines or whipping or branding. For the first few years of statehood on the frontier, such sentences seemed quite satisfactory. In 1826, Governor John Murphy reported to the legislature that the Alabama system of government did not contain "many prominent or glaring defects," and he warned the lawmakers against the "evil of too frequent legislation."[1]

The perceived perfection of one time, however, becomes the glaring error of the next, crying out to be changed and reformed, attacked as an intolerable impediment to justice and happiness. In 1836, the death penalty was narrowed in its application, and a sentence of imprisonment was established for a variety of crimes. In direct cause and effect, Governor Hugh McVay in 1838 asked the legislature to authorize the building of a penitentiary. After some bickering the legislature in 1839 gave its approval to the project and appropriated thirty thousand dollars for construction. There was argument on the location of the prison. When Wetumpka (the county seat of Elmore, located seventeen miles north of Montgomery, the state capital) was chosen, there was further acrimony on where in Wetumpka it should be built.[2] Two years and an additional fifty-three thousand dollars later, the prison was completed and turned over to the Board of State Inspectors. In 1841, William Hogan, the first warden, turned the key on the first prisoner, William Garrett, a white man charged with harboring a runaway slave. The great experiment had begun.[3]

It did not go well. For five years state officials wrestled with the hitherto unknown problems attendant to operating a prison—especially to operating one at a profit. Little went right. The problem of high prison mortality appeared early. Presiding over what came to be called the Walls, Warden Hogan not only exceeded the rules of punishment but then forced those he had maltreated to campaign actively for his reappointment. One of the wooden factory buildings was burned down, apparently by a prisoner strongly opposed to the work ethic. There were escapes and additional expenses for tighter security. These problems might well have been anticipated and therefore accepted with the equivalent of governmental good humor. But far transcending bad wardens and troublesome convicts was the fact that the prison was not paying for itself. It was losing money, an intolerable situation.

If the state could not run the prison at a profit, perhaps others could. On February 4, 1846, the legislature took a momentous step and authorized the leasing of the penitentiary for a six-year term. After paying the state inspectors and promising that the prisoners would have wholesome food and decent clothing, a lessee could decide how to utilize the premises. The separation of the prisoner from the state and its control thus began early, and it set a fearful precedent.

For six years John G. Graham served as warden and lessee of the penitentiary—the uncontested winner with his bid of five hundred dollars a year. The scanty records indicate that Graham ran a profitable operation in the manufacture of buggies, harness, saddles, and shoes. They indicate as well that the prisoners may have been cared for somewhat better than when they were under state control. The very success of the lessee revived the thought that the state was being cheated out of money it could have made for itself, but when the Graham lease expired in 1852, another six-year lease was awarded to Dr. M. G. Moore and F. Jordan. As in the earlier lease, the state allowed its fee to be paid in the form of capital improvements. Moore's manufacturing efforts lost money, and conditions among the prisoners took a decided turn for the worse. In 1858, the lease went to Dr. Ambrose Burrows. With over two hundred convicts, Burrows might have used the war years to reap a profit. But he was murdered by a convict in 1862—no testimony perhaps to the gentleness of his management.[4]

With war in progress the state assumed control of the penitentiary and named the second lessee, Dr. M. G. Moore, as the warden. With a guaranteed market in war goods, the prison probably turned a profit to the state for the first time. With cries of freedom and emancipation in the air, Northern troops arrived and "in their zeal for liberty" released all the prisoners. If there was a bright side to such action, it lay in the creation of a chance to rethink prisons and to order them on a better foundation and along more efficacious avenues.[5]

Lost opportunities, missed chances, and errors mild and vicious litter the tracks of governments as well as the dimmer footprints of individual citizens. Reconstruction in Alabama was assuredly a time of change. Large parts of an unjust past were abandoned, but they were often replaced by new practices as blind and reactionary as the old. Prison policy was one of the chips of public affairs tossed about in the ebb and flow of political maneuver and carried to its destination by the irresistible current of money and self-interest.

There was no dearth of prisoners to allow the issue of their maintenance to remain a secondary issue. The Penal Code of 1866 contained those provisions of social restraint and economic discipline remembered as the Black Codes. County courts could "cause to be hired out such as are vagrants to work in chain-gangs or otherwise, for the length of time for which they are sentenced."[6] While vagrancy had its special provisions, hard labor for the county was the punishment for a host of offenses. Those found guilty of felonies were sent to the state penitentiary, while the county jail was the destination of persons convicted of misdemeanors plus costs. Subsequently the prisoners would be leased either within the county or elsewhere.

The code made it plain that the motives behind punishment at hard labor were little removed from the labor economics of the Black Codes. Hard labor included the "hire to some other person, or corporation, to labor anywhere within the State."[7] Although the Black Codes brought counterattack, hard labor for the county—often no more justified than the codes themselves—remained an accepted and defended mode of punishment and profit.

The Convict Lease System

The next development was one of incredible viciousness. On one point Governor Robert M. Patton (a prewar Whig, elected in 1865) practiced the same philosophy as the post-Reconstruction Bourbon Democrats. Patton persuaded the legislature to reestablish the leasing of the penitentiary but stipulated that convicts must be leased outside the Walls. Blacks, Patton argued, did "not regard confinement as any punishment at all." Yet they would "surely feel the hardships of labor in iron and coal mines."[8]

Patton proceeded to lease the prisoners to Smith and McMillan—a bogus front for railroad builders and for iron and coal entrepreneurs, including such later Birmingham titans as Thomas Peters, Sam Tate, and James W. Sloss.[9] The activities of Smith and McMillan proved typical of future lease contractors. To add bitter insult to desperate injury, the firm paid only eight thousand dollars of the twenty thousand it owed the state. Beyond that, Smith and McMillan received a fifteen-thousand-dollar loan from the state that was never repaid. Patton created a leasing system devoid of any state control. It was an invitation to abuse, a "system" that killed 41 percent of its prisoners in 1870.[10]

The subsequent history of convict leasing in Alabama was tied directly to industrial expansion in the state's north-central mineral counties. Beginning in the 1870s, Birmingham and Jefferson County emerged as first among a complex of booming New South cities and counties. Once penetrated by railroads, especially the L & N under the hard-driving Milton H. Smith, the area became a major producer of coal, coke, pig iron, iron, steel, and related products. An ambitious group of Gilded Age promoters emerged to practice the Gospel of Wealth and to help define the state's political and economic policies. Both native Southerners and Northerners, they included James R. Powell, Henry F. DeBardeleben, John T. Milner, Truman H. Aldrich, and James W. Sloss.[11] Convict labor cost less than free labor, and various industrialists bid for it without flinching. The majority of convicts went to the mines, although some were leased to south Alabama lumber camps and farms.

The TCI became the area's archetypal corporation—including its leasing of convicts. Originally a Tennessee concern, the company made investments in Alabama, sent its top personnel there, and by 1892 was dominant. The mineral district's railroad promoters and industrialists formed a political alliance with the planters and merchants of the Black Belt. As one study has shown, this urban and rural Bourbon alliance was not without its differences; basically, though, it held because of a common laissez-faire philosophy in economics and conservatism in politics.[12]

As such larger developments unfolded, the state penitentiary, stripped of its valuables, became dilapidated. Republican Governor David P. Lewis (1872–1874) decided to purchase the 1,800-acre farm of Thomas Williams, located near the prison. The negotiation would allow an alternative to leasing and promised at least some profit from staple crops. The state bought the land without realizing that Williams still had at least half of it under mortgage. No sooner had that problem been settled than it was revealed that most of the purchased land was subject to flooding by the Tallapoosa River. Between flooding and insect infestation, the first cotton crop was either washed out or eaten up. The prison system of the Reconstruction era was an unending labyrinth of fraud and chicanery, a monument to inventive accounting and barefaced stealing.[13]

The years of Reconstruction did not simply kill convicts and make profits for railroad builders and iron makers. Those were lesser matters to Governor George S. Houston and his fellow Bourbon Democrats. They "redeemed" Alabama and regained political control in 1874, only to discover that the convict system did not make money for the state. Such a condition violated the Bourbon shibboleth of strict economy, which, along with honesty in government and white supremacy, was a source of strength. The Bourbons had excoriated the profligacy of Republican hegemony and had hammered home their claims of fiscal responsibility. Thus, public policy had to be changed, and Houston received accolades when he decided to lease able-bodied convicts to contractors. The Walls would be retained to house the ill and the aged. Under the promptings of J. G. Bass, Houston's nominee as warden, state inspectors were removed from contractors' work sites because, as Bass brazenly explained, they created difficulties for the contractors.[14]

Here was a system so perfectly open to exploitation (and corruption) that it might well have been the direct plan of the leasing companies—and perhaps it was. Convicts were completely under company control. The hand of the state appeared only in quarterly visits to the larger lessees by the state inspectors. From 1875 to 1878 the number of prisoners increased; the overwhelming majority of them were blacks. It was all simple to explain. One observer stated as fact that for many former slaves "freedom meant license." As a consequence, he added, impoverished southern states could not house the overflow, and the "convict lease system was a natural result; it proved profitable both to the state and to the lessee, as the latter could almost always underbid free labor."[15]

Warden Bass played his role. He energetically sent the convicts out to the turpentine stills and the coal mines. The inspectors reported on the "humanity of treatment" given by the contractors. There were signs and portents that matters might be otherwise, but these were overlooked in the glow of success that radiated from Warden Bass. In 1877, Bass paid the state treasury fourteen thousand dollars, and the next year he did even better with sixteen thousand.[16] Annual accountings embossed in black ink muffled any would-be critics. It was not without significance that George W. Cable entitled his book opposing leasing *The Silent South*.

Trouble began for the triumphant warden when Governor Houston retired and Rufus W. Cobb assumed the reins of power in 1878. Cobb was no friend of Warden Bass, and he was convinced that contracts for better than five dollars a month per head could be made. Cobb adroitly cut the warden out of the bidding process, and in the maneuvering that followed, the price was increased to six dollars and then to eight. Governor Cobb made a better arrangement for the state, but also he placed the convicts into the hands of his friends—and that, perhaps, had been the real point of contention between Cobb and Bass.[17]

The thrusts and counterthrusts between Cobb and the warden

were a mild preliminary of what was to come. In 1881, a legislative committee, with Governor Cobb's blessing, investigated the prison system. The auditors reported that, on the whole, the system was benevolent. The convicts were "kindly and humanely treated."[18] Having blessed the system, the committee castigated Warden Bass and recommended that his powers be sharply circumscribed. Bass completed his term in February 1881; among his last official acts, he signed new contracts. It was the end of the Bass era, and it was the beginning of matters both better and worse.

The time and patience of a governor have often been extended by the cries and pleas of the importunate, reminding him of their fealty to his cause. They carefully document their dire need for the comforts of a state appointment. Old files bulge with the accumulated letters of request. In this bid for place and position, few have ever surpassed the efforts of John Hollis Bankhead. Aiding Bankhead as he sought the now-vacant post of warden were the influential and prominent. From practically every county came letters extolling the abilities and virtues of the applicant.[19] Governor Cobb yielded to the pressure.

A man of force and ambition, Bankhead was born in Lamar County (then Marion) in 1842. He was self-educated, served in the Confederate army, and was wounded three times. When he was appointed warden in 1881, he had already served in the state house of representatives and senate. Bankhead was on the road to Washington, D.C.—first the House, then the Senate—as he moved within the rewarding folds of the prison system. He was warden for four years, a length of time sufficient to change the convict lease once again.[20]

When he took office, Bankhead canceled the contracts recently signed by Bass and Governor Cobb. The new warden insisted that they were illegal in their vagueness and were against the interests of the state. In their place Bankhead signed one-year contracts, a clear sign of a temporary arrangement with something else to come. It came as Bankhead sent the state medical officers to the convict mines. The medical reports painted the vilest scenes of filth, disease, and cruelty. Here was documentation that gave the lie to the reiterated claims by convict inspectors that all was well. Cast as a reformer

John Hollis Bankhead, United States congressman and senator, served as warden of the state prison system, 1881–1885. (Birmingham Public Library Archives)

and humanitarian, Bankhead watched approvingly while Alabama newspapers spread the word of evil and the need for change. Where the committee of 1881 had attacked the warden but spared the system, Bankhead arrayed the matter to show the system in its foulest light. His actions implicitly show that he wanted change. It was not yet clear whether Bankhead wanted only the purification of existing arrangements or whether he was a foe of the lease system itself.[21]

Allowing sufficient time for public agitation on the question, Bankhead announced that the system was "a disgrace to the State, a reproach to the civilization and Christian sentiment of the age, and ought to be speedily abandoned." It should be abandoned in favor of the Bankhead plan, the only way to safeguard the health of the convict miners while producing revenue for the state. All the convicts should be leased to one coal company for ten years. The company would build a prison at its mines, and the warden and other prison officials would move with the prisoners. Of course, Bankhead admitted, that would create a monopoly of prison labor (and give an immense competitive edge to the lucky company). Still, "It is not the business of the state to inquire who will be benefitted or injured by the change."[22]

The prison inspectors agreed that the convicts would fare better under the warden's direct control, but they objected to concentrating all the prisoners in mining. Governor Cobb opposed every part of Bankhead's plan and pointed out quite correctly that Bankhead himself could have stopped the practices that had been exposed.[23]

The state legislature, under the newly elected Governor Edward Asbury O'Neal, seemed to deny Bankhead on essential matters. Contracts were limited to five years, and not more than two hundred prisoners could be leased to one company. The Walls would not be sold, but Bankhead was free to move the major part of the convicts into coal mining. On the matter of county convicts—a prickly issue on which Bankhead had remained silent—the legislature spoke the language of local interest. The body decreed that perpetrators of misdemeanors "shall be employed or hired in the county convicted, unless in the opinion of the county it requires that they be hired outside the county."[24] The practice, made legal in 1866, was perpetuated.

With a new board of convict inspectors taking office on March 1, 1883, Bankhead began to implement the great plan. The warden advertised for bids, and on April 1, it was announced that the Pratt Coal Company would be awarded the legal maximum of two hundred convicts. Comer and McCurdy, who mined on Pratt land, would get two hundred, and J. F. B. Jackson would receive one hundred for agricultural work. It appeared that the large mining interests of John T. Milner were the loser in the battle of the bids. Once again, appearances bore no relation to the realities. The reasons are instructive.

Before the contracts were even signed, critics had observed that Pratt and Comer and McCurdy were in fact the same company—either the same or so collusive as to destroy the meaning of "separate entity." Any legal niceties were handled by having "competent legal authority" obligingly rule that the companies were separate. The effect was to grant four hundred state convicts to Pratt. County convicts were divided between Comer and McCurdy, Milner, and the mine operations at Coalburg in an agreement that gave each company its own counties to draw on. But Pratt got county convicts from Comer and McCurdy, and Milner controlled the Coalburg mine. Pratt and Milner thus had a monopoly on both state and county convicts.[25] The rankest collusion had governed Bankhead's contracts. Only the innocent and the naive could conclude that it was a surprise to the warden.

Warden Bankhead moved his official residence to the Pratt mines in Jefferson County and began to supervise the building of new prisons to be paid for by the companies. Even as Bankhead saw his plan take on reality (and he proudly spread the word of his zinc-lined bathtubs ninety feet long), the latent but countervailing force of the inspectors began to stir.

Throughout 1883, the inspectors, led by the strong figure of R. H. Dawson, a conscientious and self-taught expert on coal mining, launched investigations that showed the continuation of filthy conditions and endless whipping. Their reports detailed the death tolls that coal mining exacted. The investigations and reports were attacks on the system, but they were implicitly attacks on Bankhead himself. In areas of their jurisdiction the inspectors took action. They researched the complex problem of convict classification and, in the face of company opposition, formulated their own work quotas for the prisoners.[26] In public reports the inspectors indicted Bankhead for condoning excessive punishment, established his inefficient handling of financial matters, and argued against the employment of convicts in coal mining. It was a fair indication of their views that they recommended abolishing the office of warden.[27]

Like Warden Bass before him, Bankhead was now brought low. On February 17, 1885, the state legislature enacted the Coleman law, authored by Representative Augustus A. Coleman of Greensboro, Hale County. The measure abolished the office of warden and placed authority in the hands of the president of the board of convict inspectors and his two colleagues. County convicts, if used in coal mining, were to be inspected on the same basis as state prisoners—a long overdue reform for the worst part of the penal system.[28]

Bankhead had used reform. He used it callously and hypocritically in his own interests and to the profit of the coal companies. His claims of reestablishing state control sounded good and were literally true. Yet Bankhead's "achievement" obscured his real accomplishment: the system itself had been delivered into bondage. Not once—not one time—did Bankhead use the powers of his office to right the very wrongs he preached against. Bass was a crude exploiter; Bankhead used good to produce evil—an ethical conundrum that historians luckily need not solve.

It was left to inspectors Dawson, A. T. Henley, and W. D. Lee to manage a system they believed to be inherently wrong. Bankhead's contracts would not expire until 1888. The goal would have to be better conditions in the same exploitive work. Now the inspectors regulated procedures with a host of new forms designed to guarantee oversight and responsibility. The leading edge of bureaucracy attacked the Alabama penal system. On numerous occasions

Representative Augustus A. Coleman, Hale County, in 1885 introduced the law ending the era of powerful state wardens by placing prison administration under a board of inspectors. (Alabama Department of Archives and History)

the inspectors intervened quickly in cases of brutality and successfully fought to uphold their authority to oversee the county convicts.[29] With the state public health officials, they kept a watchful eye on sanitation and disease, and they allowed Julia Tutwiler, the determined woman reformer, to start her convict schools. In the face of national attacks on the Alabama system, Dawson argued that state supervision and control had changed the crudities of the lease system. But all the inspectors agreed that coal mining was a destructive occupation, and they strongly supported the appointment of a commission to reform the system.[30]

Historians find it hard to resist associations and connections based on simultaneity. If a reform movement is in progress and dominates the moment, they give to it the birthright for the changes that may have been sired in earlier times of deep conservatism. The decision in Alabama to withdraw all convicts from coal mining was first made during 1890–1893. Agrarian reformers, spurred by a variety of motivations, attacked the ruling Democratic party and its strong ties with business interests. Populism's forerunners, the Greenbackers and Independents, had condemned the convict lease system in their platforms of the 1880s. Then the Jeffersonian Democrat (and latter-day Populist) Reuben F. Kolb demanded an end to the system in 1892. Abolition of leasing was compatible with the radicalism of the day, but its positive origins were with the bureaucrats of the board of inspectors. Incongruously, they had support and initial implementation from Governor Thomas G. Jones, a paragon of antilabor, probusiness Bourbonism. Things were not as simple as they seemed.[31]

In February 1891, the Alabama legislature passed an act establishing a commission to investigate alternatives to leasing prisoners. The governor, the board of inspectors, and one person appointed by the governor were instructed to examine the possibilities of working all the convicts "on State account." They were to prepare a bill "creating a new and complete convict system for the State of Alabama."[32] The commission concluded that public opinion was opposed to the lease system and to working convicts in the mines "under any system." Mining was considered to be an "internal improvement" from which the state was constitutionally barred. Other forms of state account would cost enormous sums of money in capital investment. The commission's answer was to recommend continuing the present contracts. The money earned would be placed in a special fund that would eventually pay for a new system. At the same time the commission suggested a new form of governance. To remove the system "as far as possible from partisan control," a nine-member Board of Managers, representing every congressional district, would supervise the three inspectors.[33]

In November 1892, Governor Jones recommended the plan to the legislature and stressed the impossibility of doing more for the present.[34] In February 1893, the legislature did its duty and passed an act creating a new system. It provided that all convicts must be removed from coal mining by January 1, 1895, "if it can be done without detriment to the financial interests of the State." Contracts for convict labor could be made, but prisoners had to remain under the direct control of the state. Hard labor could be used for production and profit "or for the purpose of intellectual training and instruction, or partly for one and partly for the other."[35]

When the contract with the Tennessee Coal, Iron, and Railroad Company ran out, it was the intent of the law that all convicts would be removed from the mines; they would be transferred to the penitentiary "or some prison" as rapidly as possible. County convicts would share in this incredible day of emancipation; they, too, would be removed from the mines after January 1, 1895, and their sentences for costs were not to exceed six months.[36] The evils of the coal mine lease system were about to end. The system was ready to occupy new ground and march out in new directions of rehabilitation and compassion.

The good intentions and the legislation lasted for one year. The nine managers (with Dawson now superintendent) decided that the old prison at Wetumpka was useless as the foundation for a new system. After suitable investigation the managers purchased 4,508 acres of land at Speigner in Elmore County. The managers particularly wanted to build a women's prison. They were acutely aware that, if the state did not have prisons, it remained at the mercy of the lease system.[37]

The panic of 1893 applied the strongest vetoes to new directions; also, a committee of nine was perhaps less than efficient in running a prison system. Nevertheless, in 1895 the legislature terminated the Board of Managers and restored the system to the control of the Board of Inspectors—Dawson, Henley, and Lee. In truth, no prisoners had been taken out of the mines. To remove them meant prisons to put them in; prisons meant money, and the state treasury was empty. Dawson sought to put a good face on the matter and argued that the demise of the managers did not mean that its "plans and policies were not wise." Those plans should be followed and "enlarged upon" as soon as possible. When the lease expired, the state should be "as nearly independent of contractors as possible."[38]

Dawson's hopes were quickly shattered. A small cotton mill was built at Speigner from convict department funds. Even so, with no capital available to enlarge the prison factory, it seemed certain that the state would have to lease to the mining companies once again. An aura of failure hung over the system. The night school program at the Walls was a failure (and a success among the black convicts at the mines); the death toll among the convicts rose to 14 percent for state prisoners and probably as high as 24 percent for county convicts.[39] The system stumbled in its policies, and the thrust for reform, once thwarted, sagged into frustration and impotence.

At this juncture a special committee of the legislature was appointed. Perhaps something could be salvaged, or at the least, a workable program might be developed. The committee's grasp of the broader issues was not well displayed in its conclusion:

> If convicts were reliable and good working men prior to their conviction, then their work as convicts would not add to their aggregate of labor products; but such is not the case. There can be no question that convicts produce more after than before conviction; the total production of the criminal classes is much

greater after conviction than before; were it otherwise, there would be very little crime, and consequently, very few convicts.[40]

On specific issues the committee did much better. It found no satisfaction in reporting that convicts made up 13.23 percent of all coal miners in Alabama and 26 percent of all in Jefferson County. It viewed such concentrations as an unfair imposition on free labor and recommended using only the most able quarter of the prisoners in mining. The committee thought TCI unlikely to renew the lease in 1898 and suggested that the state consider mining on its own account.[41]

Committees of the legislature had been one of the few avenues for investigation, exposé, and reform. The committee of 1897 had viewed the Dawson policies with favor. It felt the pressures of anti-lease opinion, and it saw state account as an acceptable substitute. But the ebb and flow of prison policy was not affected entirely by the rhythm of ideas. Hard interests and material motives were more likely to prove the winners. In this case, the election of 1896, where classic versions of liberalism and conservatism met in direct combat, confirmed the route of convict policy. Joseph F. Johnston, president of the Alabama National Bank, cofounder of the Sloss Company (a major user of convict labor in its mines), was elected governor. Johnston took pride in the support given him by "the plain people" and, in fact, was viewed as a neo-Populist. He was also described as a Progressive. If Johnston represented a transition from Populism to Progressivism, he was in practice a thinly disguised Bourbon. Johnston gave his support to business interests—especially including his own.[42]

With the Johnston administration, Alabama began sewing its patch to Progressivism's highly irregular quilt. If there was a consensus that equated Progressivism with reform, contemporaries differed widely on its implementation. Historians have assigned shadings of origin, motivation, commitment, exclusivity, geography, and significance to the movement.[43] Operating within the confines of white supremacy, Southern Progressivism was highly selective. It was conservative, middle class, and urban—decidedly not radical. Yet, in Alabama, Progressivism had an agrarian flavor and was influenced by the programs of Kolb and the Populists. Besides Johnston, Governors William D. Jelks, Russell M. Cunningham (acting), Braxton B. Comer, and Emmet O'Neal would face the issues: education, prohibition, railroad regulation, taxation, democratizing government, and the many facets of social welfare.

Alabama adopted a new constitution in 1901 whose chief purpose—one that it achieved—was disfranchising the black voter. If the document seemed an automatic barrier to real reform, there were those who argued that it cleared the air. Now, they postulated, whites could address the real issues. There would be no need to concern themselves with the political nightmare—born during Reconstruction and born again with Populism—of how blacks would vote.[44]

Penal reform, even had there been no tragedy at the Banner Mine, was one of the "real issues." It was a natural area of concern for the Progressives and reformers. The challenge was to rectify a system that a modern historian described as one of "cruelty and brutality" and another as "incontrovertible evidence of the New South's moral failure."[45] Johnston and the chief executives who followed him could not ignore the issue.

Governor Johnston named Sydeman B. Trapp to the presidency of the Board of Convict Inspectors, and Dawson and his care-worn colleagues finally left the system. Where Dawson had moved toward state account, Trapp recommended selling the mill at Speigner and disposing of all the prisoners on the state farms "by lease or otherwise." Trapp was proud that the United States Industrial Commission cited Alabama as the operator of the most profitable lease system in the nation—net earnings of $188,533.60 on a gross of $325,196.10.[46]

The results of Johnston's "businesslike administration" of the prison system were appalling. Fraud and corruption always ate away at the edges of the system, but the actions of Trapp and Johnston rivaled the most fraudulent activities of Reconstruction. TCI and Sloss Iron and Steel received five-year contracts substantially lower than those paid in the earlier four-year period. The negotiation of the contracts was marked by chicanery, and they were not prepared by the attorney general, as required by law.

Later contracts were made without advertising for bids; ten-dollar-a-month convicts were billed at five dollars, and five-dollar convicts at three. President Trapp excused these "irregularities" with the odd defense "that it was all done with the sanction of the Governor." There were few rules and regulations of the department that were not broken, and President Trapp's "table . . . seems to have been supplied with meats and vegetables" and his house with "wood, corn, and hay" from the prison farm. Trapp and Governor Johnston personally handled all purchases for the system. There was no competitive bidding, and the business went to two or three firms.[47]

A joint committee of the legislature probed the Johnston administration's excesses and was rightly struck by the inconsistencies of Alabama policy. A cotton mill had been built and then allowed to stand idle; the lease system itself was to be abolished but continued instead without a break. The inauguration of a far-sighted policy by one administration, said the committee, "and the total abandonment of it by the next cannot but result in harm to the State. Some fixed policy should be determined and carried out."[48]

The committee was correct, but the question remained as to what the "fixed policy" ought to be. J. M. Carmichael took over as president of the board of inspectors on March 1, 1901. He simply ignored a contract made the day before he took office (but not approved by the governor) and proceeded to make new ones. The contracts showed a strong trend toward small leases for agricultural purposes, and they were made on better terms than the old.[49] The cotton mill, finally closed by Trapp, was leased to a private company, with the state meeting all convict expenses.

In December 1901, the board made new contracts with the coal companies, with a change in form. TCI and Sloss-Sheffield agreed to pay the state on a basis of per ton of coal mined. The entire housing, feeding, and clothing of the prisoners would now remain in the state's hands. The state would work its prisoners and sell the coal they mined to the companies. The new arrangement answered the charges of corporate exploitation, but it shifted the pecuniary motives for exploitation to the state.[50] Two years later, and with a different president, the system was judged to be financially favorable and morally deficient. That had been the conclusion of earlier decades.

The long-time separation of county and state convicts that went back to the "hard labor law" of 1866 (with additional legislation later) was revised in 1907. Convictions meant fees for county officials, and leasing those convicted brought in even more money. Officials in the capital decried the system, since it meant less money for the state. Reformers, whether guised as Populists, Progressives, or simply outraged citizens, opposed county leasing because there was less control. As bad as state administration was, that of the counties was worse. In any event, a special session of the legislature in 1907 made county convicts subject to the same regulations as prisoners of the state. The measure was billed as a Progressive victory of the Comer administration. Actually, the counties were permitted to continue leasing convicts. The counties still used the prisoners as a major source of revenue; the convicts themselves derived few benefits—those who survived the Banner Mine explosion could attest to that.[51]

The adhesive tenacity of monetary gain—for state and for company—had overridden every effort to change the system. Indeed a system is closest to guaranteeing perpetuity whenever the profit motive is on its side. So the state lease system and the evil county fee system survived into the twentieth century. As might be expected, free labor in general and organized labor in particular objected to competing with state and county convicts.

Coal extraction, known early in Alabama history, expanded during the Civil War and became systematic during Reconstruction as mining companies were formed. Not even the crippling effects of the depression of 1873 stopped the industry's growth: in 1878 Alabama produced 224,000 tons of coal, a total that rose to 8,504,327 tons in 1900. By 1911 Alabama miners produced 15,021,421 tons.[52] Most of the mines were located in Jefferson, Walker, and Bibb counties, although coal was also mined in Shelby, Tuscaloosa, St. Clair, and Talladega counties. Both white and black miners were employed; as time passed, a mix of ethnic migrants from both the northern and the eastern and southern parts of Europe entered the work force. Also, convicts were used on an ever-increasing scale.

Beginning in the 1870s, the Knights of Labor organized some of the miners. Although the Knights existed until the end of the century, the organization's lack of leadership or central control, its contradictory programs, and its all-encompassing concept of membership cost it vitality. The white Knights were usually opposed to cooperation with black members on issues of mutual concern.[53]

The miners sought further self-protection. They affiliated with short-lived political parties that advocated inflation, benefits for industrial labor, and an end to Bourbon control. Invariably, the issue of white supremacy prevailed, and the Democrats won. Without exception, the Independent, Labor, and Greenbacker parties of the 1870s and 1880s, as well as the discredited Republicans, strongly opposed the convict lease system. The free laborers took another tack in 1885 by forming the Miners' Union and Anti-Convict League. It foundered on the issues of race, the use of strikes, and the admission of nonminers to membership. Although some Alabama miners affiliated with national unions in the late 1880s, nothing proved permanent until the appearance of the United Mine Workers of America (UMW), organized in Ohio in 1890.[54]

The UMW entered the state in the early 1890s, but the United Mine Workers of Alabama was formed independently in 1893. Not until the late 1890s was there a permanent merger, and Alabama became officially known as District 20. A strike by the UMW in 1890 failed to win union recognition. Defeat came because Alabama's operators utilized convicts and imported "scab" laborers. Organized labor scored a minor victory in 1891, when the legislature enacted a mine inspection law. Every three months an inspector of mines

($1,500 annual salary) was to check all mines employing twenty or more workers. Safety requirements were spelled out, and violations would be heard by a board of examiners—the inspector and two mining engineers. The board could recommend court action. Although the law contained numerous flaws and loopholes, it was a beginning and a victory for labor.[55]

Three years later, labor lost a bitter and violent encounter with management. A strike began in April 1894, when TCI, as well as other operators who took their cue from the industry's leader, declined to recognize the UMW of Alabama as the miners' bargaining agent. Instead, the operators attempted to make separate contracts with their employees. The strike involved over eight thousand men, and before it ended in August, Governor Thomas G. Jones had called out the state troops. Ostensibly the militia was present to preserve order, but actually the troops protected the owners' property. Jones also employed Pinkerton detectives. When the strike failed, the operators emerged more powerful than ever. The UMW of Alabama barely survived and had to begin a slow rebuilding process. Governor Jones called the strike "the most formidable and threatening commotion in the history of this State in times of peace." Such statements prompted organized labor to support Kolb and the Populists in their unsuccessful struggles against the Bourbons.[56]

Industrial peace after 1896 enabled the UMW of Alabama to go national and to increase its membership (especially among the blacks). Still, in 1900 the union's most basic demand—ending the convict lease system—was unrealized. The operators, who organized the Alabama Coal Operators Association (ACOA) in 1900, accepted the UMW as the bargaining agent for the miners and agreed to settle disputes by arbitration. The test came in 1903 when a contract dispute was settled in favor of labor by an arbitration board. A new day in labor relations seemed to have arrived. John Mitchell, president of the national UMW, said proudly that Alabama had set an example for the nation to follow.[57]

Outer trappings of UMW strength, no less than the supposed rapport with management, were misleading and transitory. The facade was soon shattered. A state antiboycott law in 1903 crippled the strike as a weapon. Next, a series of mine disasters (though none so horrible as that at the Banner) made a mockery of the mine inspection law. Nor was there any effective legislation that established employers' liability. It came as no surprise when the coal companies began firing union men and refusing to negotiate contracts with the UMW. Edward Flynn, UMW president of District 20, and William R. Fairley, Alabama's representative on the UMW's International Executive Committee, obtained only a taunting promise from the operators: wages would be reduced.

Reinforcing management in its intransigence, Acting Governor Russell M. Cunningham refused to remove the convicts from the mines. Why talk contracts with the UMW? In desperation the UMW initiated a wholesale strike in July 1904, the first major walkout since 1894. When Flynn, Fairley, and other UMW officials offered to arbitrate, the operators refused. Although the union obtained funds from the national UMW, financial aid could not be furnished indefinitely. The beleaguered union next turned to politics.

In the primary elections of 1906, the UMW backed Jesse F. Stallings, a Progressive Democrat who was prolabor, for the United States Senate. The field was large, and Stallings lost badly as the veteran John Tyler Morgan was reelected. Morgan died within a year and was replaced by John Hollis Bankhead, former state warden. In the governor's race, a heralded Progressive, Braxton Bragg Comer, who was also a wealthy Birmingham industrialist, won easily. It was ironic that Comer's major investments were in textiles. Faint hopes by labor that Comer would favor its cause proved disastrously false. Considering the obstacles it faced, the UMW displayed amazing tenacity. The strike dragged on for twenty-six months before it ended in August 1906. The result for the UMW was bankruptcy, total defeat, and a decline in membership from eleven thousand to less than four thousand.[58]

A deterioration in leadership followed (J. R. Kennamer, who succeeded Flynn, was ill), as well as a severe decline in morale within the UMW's ranks. The coal operators moved to crush the union.

UMW men were fired. Workers who subscribed to the national *Indianapolis United Mine Workers Journal* or to the local *Birmingham Labor Advocate* were subject to discharge. Union leaders were bluntly informed that they had to accept the wage scales set by the operators or else nonunion contracts would follow. As usual, management dealt from strength: convicts and, when needed, imported scab labor were ever-present tools of power. The union had to contend with an unsympathetic state administration and with a press corps strongly favorable to management.

To retain its hard-earned role as the voice for the miners and, no less, to avoid total destruction, the UMW was forced to act. It could not permit the emasculation of having ownership dictate wage scales and all the details of contracts. Although greatly weakened, the UMW called a strike. The almost quixotic strategy was to regain status by arranging an equitable settlement through arbitration. Union men walked off their jobs on July 6, 1908.

The anticipated arbitration never materialized. When miners were evicted from company homes, the UMW leased land and erected tent cities as temporary refuges. Tension flared into violence, and the operators reactivated the Alabama Coal Operators Association. Governor Comer ordered state troops into the area, and in the interests of hygiene and sanitation, the militia destroyed the tent cities. Newspapers that had habitually branded the UMW as socialistic and communistic now added racism, with telling effect. Black union men maintained the strike despite entreaties from a black newspaper in Birmingham to return to work. Noting that a majority of the strikers were blacks, white journalists raised charges that the UMW was promoting social equality. The union's national journal replied that, because Jefferson County's black miners refused to become strikebreakers, all UMW members felt the wrath of the coal operators, Birmingham's merchants, and state government.[59] (Blacks actually represented a majority of all miners—union and unorganized—although none held a position of real power within the UMW.)

Its treasury depleted by the strike of 1904–1906, the UMW was unable to continue the struggle. By the end of August, less than two months after the walkout began, the union capitulated. The UMW had suffered not just defeat but annihilation. By 1909 there were barely seven hundred members of the UMW left in Alabama.[60] For the coal and iron and steel operators, the open shop was now an unchallenged reality. Union labor could never be powerful as long as the convicts, most of them black, were an ever-replenishing work force for management. As President George G. Crawford of TCI stated in 1911, "The chief inducement for the hiring of convicts was the certainty of a supply of coal for our manufacturing operations in the contingency of labor troubles."[61] Such was the story and such were the reasons why convicts still worked in the Alabama coal mines in 1911.

More specifically, the convicts who died in the flame and the blast of the Banner Mine were the victims of a system. A newspaper thus reported a few days after the Banner tragedy: "The long expected has happened. . . . This explosion has been predicted by every informed miner in Alabama. . . . The cause of the explosion is easy to locate in the mind of any man who understands how the convicts are handled."[62]

3
"The Facts Should Be Known"

The public reaction to the explosion at the Banner Mine was swift and came from disparate elements and interests. It was personal or corporate or journalistic. The state legislature, completing its quota of new laws and preparing for adjournment, now heated up with new issues and new debate. The Banner explosion raised together two interest-laden issues: the volatile question of working convicts in the coal mines and the urgent matter of mine safety. In each one the public conscience did battle with cost cutting and economics. The force and thrust of the Banner was to illuminate in one physical act two areas of social concern.

Mine safety and the convict lease system were publicly and privately discussed, often in the same context, but the state legislature considered the matters in separate channels of debate. Even under the spur of aroused public opinion, it was much easier to deal with legislation revising mining laws, tightening regulations, and upgrading safety standards than to tackle the real issues of the convict lease system. The decision to see mine safety as the signal lesson of the Banner was made almost without question, a "practical" decision if one hoped for any action in the last few days of the legislative session.

It was natural that organized labor in Alabama would join the lists of argument, its old stands now given new meaning by the Banner explosion. J. L. Clemo, secretary-treasurer of District 20, United Mine Workers, Birmingham, acted as the immediate spokesman for the national union. Clemo forwarded resolutions adopted by the International Executive Board of the UMW at Indianapolis to the governor, each member of the state legislature, and newspapers published in the mining counties. The resolutions extended sympathy to the families of the victims and condemned the convict lease system that exposed inexperienced convicts to a dangerous work situation and that placed them in direct competition with free labor. The system, said the UMW, exalted profit over human life. A law abolishing the convict lease system should be passed at once.[1]

In Indianapolis, the *United Mine Workers Journal* reported, "From that corporation ridden state of Alabama again comes the news of a shocking horror that challenges description." The UMW had reason for its strong demands and its abiding bitterness. Some Alabama miners remembered the lost causes of 1894, the grinding years that followed, and the humiliating defeats of 1904 and 1908. Knowledge of its impotency sharpened UMW sentiment. As one correspondent wrote to Governor O'Neal, "I am sorry to say that today in this State the United Mine Workers Union is only a name with officers getting their commissions and salary from National Headquarters . . . with the hopes that the membership which Gov. Comer killed during his administration will be resurrected someday." Another unbowed union man looked forward to the time when he could shake hands with a UMW organizer who would help relieve Alabama miners "of the oppression under which they are now compelled to exist."[2]

The Alabama Federation of Labor, organized September 1, 1900, and affiliated with the American Federation of Labor, supported the UMW's position. The Alabama organization issued a statement paraphrasing the sentiments of the UMW Executive Board. On the local level, the Birmingham Trades Council, mouthpiece for several city-based unions, denounced putting untrained convicts into unsafe mines and condemned the state legislature for not passing a strong mine safety law.[3] One newspaper editor, sympathetic to organized labor's plight, summed up the recent past: "Men who toil in these mines unionize and ask for better pay, for better conditions. They are met with scorn by the mine owners. They strike. The state militia is called out. They are evicted from their homes. They secure tents, in which to place their wives and little ones. These tents are cut down or they are forced to vacate even this frail shelter—yet this is THE FREE country. God, what a misnoma *[sic]*."[4]

Neither Governor O'Neal nor the legislature had much to fear from the pressure of organized labor. However, in a situation where public opinion was so often the product of the press, the politicians found it impossible to avoid newspaper reaction. Editors, with all their differing interests and attitudes, were the manifest conscience of the public, the closest thing there was to a collective will. Much of the newspaper reaction to the Banner disaster was an attack on the status quo.

In Florida, a state that still retained convict leasing, the Banner tragedy was front-page news in the *Jacksonville Times-Union* for four days. It received close coverage in the *Atlanta Constitution*, the *Washington Post*, the *New York Times*, and the *New York Tribune*. The event was reported to British readers by the *Times* of London and to Californians by the *San Francisco Chronicle*. In a caustic editorial the *St. Louis Post-Dispatch* predicted that accidents such as that of the Banner "will stop as soon as it costs less to save life from such horrors [than] to waste it."[5]

With interest so often a function of propinquity, state journals provided even more intensive coverage. The response of Birmingham's black newspapers cannot be measured because their issues of April 1911 no longer exist. State and national black journals were largely dependent on wire service reports. No newspaper responded more strongly or eloquently than the *Bessemer Standard*, published in an industrial city just west of Birmingham. In an angry editorial, the *Standard* declared:

> The state of Alabama is responsible for the deaths of those convicts in the coal mines on last Saturday—yes, and for the deaths of many other miners in the blackness of mines and convict camp, of which the world knows nothing. Alabama's convict system is a disgrace to humanity, to the nation, to the State, and to every man, woman and child in it. How long will it be before the criminal loss of life as that of Saturday causes

> us to halt in our greed and avarice and selfishness. The rest of the world knows that when a man is criminal or vicious it is because he is diseased in body or mind, and so he merits pity and medical assistance. Alabama enslaves her convicts and consigns them to a dangerous and vile existence, that is happiest when it terminates in the rigor of death.[6]

Various newspapers noted that the fate of the convict miners hardly corresponded to the nature of their crimes. Alabama's convicts could be and were sent to the coal mines for such offenses as public drunkenness or profanity, riding trains illegally, vagrancy, and violating Sunday "blue" laws. A visitor to the Banner Mine in 1911 was confronted by a young black prisoner who begged to be "bought out." The prisoner's crime had resulted in a fine of one dollar. The problem was that he could not pay the seventy-five dollars in costs. He had worked at the Banner long enough to earn fifty-five dollars by extra labor. If the visitor would pay the remaining twenty dollars, the miner said, "I'll work for you as long as you say. . . . Do buy me out, Sah, please do."[7]

The *Tuscaloosa News* first concluded that it was better for criminals to die than for free miners. It revised this thought when it considered that none of the dead convicts had committed a capital crime. "Their misdeeds were of a minor nature, relatively speaking," and the men were unwilling laborers in an activity where enough free miners were available to take the risks. The *News* concluded, "That such a thing should happen is a blot on the name of justice." It all showed "the utter wretchedness of our convict laws and demands that in the name of right and humanity they be revised." Another editor wrote that the men were only guilty of "crap shooting, perhaps concealed weapons, violations of prohibition laws and like offenses." With finer rhetoric than logic, the writer concluded that they were "condemned to death, and that too, without time for preparation, a privilege granted to the felon before the springing of the death trap."[8]

Continuing the assault, the *Birmingham Labor Advocate* criticized the state's demand for punishment that bore no relation to the crime.

"Our vocabulary is too limited to express our feelings," its editor wrote, adding, "when a convict is killed in one of these mines it is simply judicial murder." Another labor editor's vocabulary was not limited. "This dangerous work was forced upon the unfortunate victims," he wrote. "They had no choice but to go wherever they were driven, and do whatever they were bidden or be scourged with a lash pitiless and keen."[9]

A Talladega editor charged that, because Alabama's laws sent convicts to the mines for trivial offenses, the state was responsible for the Banner tragedy. "These men had done nothing worthy of the death sentence," the editor of the *Jackson South Alabamian* noted, "and criminals though they were, they had a right to live." *Howle's Iconoclast*, an uninhibited Birmingham newspaper whose motto was "Stone Blind to Everything but Right and Justice," cut directly to the center of the controversy: the state wanted the money and the mine owners wanted the cheap labor; thus, "the State of Alabama is to blame for the deaths of every one of those lost lives at Banner." A journal in another mining county declared simply, "Alabama is responsible for it."[10]

In Montgomery an independent editor responded angrily: "The soul sickens . . . at the thought of a great sovereign state placing its prisoners in such jeopardy in order that its treasury may grow. . . . The hiring out of convicts by the state is wrong from beginning to end." In Escambia County, a piney woods region far to the south that bordered Florida, the tragedy did not go unnoticed. "Is it human to take a man who has never seen the inside of a mine and put him to work in one?" a local editor asked. He answered his query in the negative, observing that the Escambia prisoners "were sent to their death for stealing or some other minor offense." In Alabama "you can't hang a man . . . for grand larceny, yet you can send him to the mines, where he contracts tuberculosis, which is ten times as bad on the victim." Six prisoners from Escambia had worked at the Pratt Consolidated mine. All were killed, and there might have been more if the next shipment of prisoners had arrived.[11]

Baldwin, a Gulf Coast county bordering Escambia, took comfort in its policy of keeping convicts at home. The system was imperfect, but as a local journalist remarked, in Baldwin County "the authorities . . . do the best they can under the existing laws and hire out the convicts on a farm."[12]

Some editors gave specific examples to make their point. Ed Causey, a white man convicted in Calhoun County, was one of the Banner's victims. For the crime of stealing whiskey, Causey was forced to work out the costs of his trial, and the county sentenced him to eighteen months' labor for Pratt Consolidated. The newspaper did not point it out, but if Causey could have paid his fine and court costs, he would not have gone to prison at all. Money was a cushion against the rougher spots of life. Charles H. Greer, editor of the *Marion Standard* in the Black Belt county of Perry, remarked that "several negroes from this section convicted of running blind tigers were caught in the Banner mine explosion. That is a pretty tight penalty to pay for selling booze."[13]

A number of commentators were struck by a major anomaly of the disaster. When previous explosions had wracked the mines at Palos and Mulga, a host of crying wives and family members had haunted the shaft head waiting for any word. At the Banner there had been an unnatural silence as the bodies were brought to the surface.[14] There were no grieving families, loud in their lamentations and waiting in uncertain outcome to claim their dead.

Some of the Banner's bodies did not have far to go. Free miners Lee Jones, white, and David Wing, black, were buried at Linn's Crossing, a short distance from the mine. John Wright, serving a sentence for assault, was interred at Sayre, his home, only three miles from the Banner. Most of the dead, however, were far from home. Their bodies were sent to the counties where they were convicted, but only if the next of kin could be located. Many remained, buried in the prison cemetery—a ditch on the wild slopes around the Banner.[15]

There was an incident of macabre boosterism, as a Birmingham paper approvingly reported the visit of one Austin A. Breed of the Crane-Breed Manufacturing Company of Cincinnati, Ohio, specialists in coffin production. Breed was considering a branch office in Birmingham, and he "believed at a glance that this [area] was a good field for a factory."[16]

Beyond the general condemnation of the relationship between the lease system and mine accidents, the press was involved with the question of what caused the explosion. A Birmingham paper declared that "a rigid examination is demanded . . . to place the blame where it belongs, . . . the facts should be known, and the responsibility located." According to one trusting newspaper, "Governor O'Neal is having a thorough examination made, and if someone is to blame that person is going to suffer." The *Birmingham Ledger* editorialized that "up to this time it is impossible to fix the blame," and the rival *Birmingham Age-Herald* echoed the sentiment that "it is too soon to say what caused the terrible explosion." Frank V. Evans, editor of the *Jasper Walker County News*, retorted: "The Age-Herald says it's too soon to learn the cause of the Banner mines disaster. Widows of the victims think it is too late and others agree with them."[17]

One line of thought, well known to more modern times, offered an easy escapism. The general citizenry, society itself, was at fault. Since the convict inspectors issued annual reports (in fact, in 1911, the reports were being published once every four years), a concerned and responsible public should have consulted the published statements. If, on becoming more informed, the people (that amorphous creature) saw that action was needed, pressure should have been brought and change implemented. This endless refrain of democratic life tried to elevate the obvious to the stature of an explanation. Julia Tutwiler—prominent educator, social reformer, and vociferous opponent of convict leasing since the 1880s—pointed out the damning reality of the situation. "I fear our prison camps would have shocked even the Turk," she wrote, with acceptable error. "I feel ashamed to think how far in the Middle Ages we are in the South." She lamented that so many convicts "met a cruel death . . . at the Banner mine"; while reports were certainly issued, she wondered whether any Alabamians except the governor actually read them.[18] Cynics wondered if the governor did.

The most meaningful explanations would be those that sought the physical basis for the explosion: the report of the coroner's jury, those of the state mine inspectors, and the findings of the United States Bureau of Mines (if Dr. Holmes ever released its report). Even so, these did not preclude the offerings of other experts and observers. Some individuals argued that the disaster could have been avoided if existing state laws had been enforced. John de B. Hooper, responding to a request for advice from Governor O'Neal, refused to be specific but remarked, "I can say this, that had I been occupying the position of chief [inspector], there would only be a remote probability of such things happening.[19]

Dr. R. P. Huger, a Democrat and leading citizen of Anniston, a quintessential city of the New South, was less self-serving and more direct: "It appears to me that the mine inspectors who have charge of the Banner Mine are either incompetent or ignorant." Huger complained that "we have such a recurrence of disasters in the mines of Alabama that they are no longer accidents, but must be attributed to some laxity of inspection or the fault of some man who may be criminally careless or ignorant." One man, with long experience in mining, inspected the Banner after the explosion and saw that the brattices had been blown down. He argued that, if they had been properly installed, rescue teams could have descended to the bottom of the mine in thirty minutes.[20]

Dr. E. Stagg Whitin, general secretary of the National Committee on Prison Labor, followed events in Alabama from his headquarters in New York City. Even at that distance his views were far more sound and relevant than those of many local analysts. Dr. Whitin carried on a brisk correspondence with Governor O'Neal and with John D. McNeel, the governor's private secretary.[21] "It seems likely," Whitin wrote the governor, "that this horrible accident would have been impossible if the section of the mine where the accident occurred had really been investigated by the fire boss." Whitin was convinced that "prevention is what is needed . . . for the conflicting reports by inspectors always disguise the real situation." The *Gadsden Evening Journal*, the voice for Etowah County's burgeoning industrial city, unknowingly agreed with the New Yorker. Pointing to existing statutes, the paper wondered why they were not rigidly enforced. The answer, its editor said, was that "laws for the protection of property are much more apt to be rigidly maintained than those for the preservation of life." It was the oldest of problems: there was no lack of laws; there was a lack of enforcement.[22]

There was specific criticism and blame for the Pratt Consolidated Company. The corporation, as revealed in newspapers, court cases, public statements, and official documents, did not hesitate to defend itself. The strategies its officials debated and adopted are unavailable because the company's records have been discarded or destroyed through the years.[23] The company also found support among certain rural and urban newspapers. A small-town Black Belt journal, the *Linden Democrat-Reporter*, argued that "science has not yet reached the point where these fearful mine holocausts can be avoided but to the credit of the mine owners be it said every method known to science is being used to protect the lives of the miners." The *Birmingham Age-Herald* also considered the owners to be blameless and brought out the frequently repeated accolade that the Banner was "a model mine." In fact, said the *Age-Herald*, "The operating officials, next to the families of the dead, will have the sympathy of the public." Scoffing at such defenses, another editor responded with sarcasm: "By reading the reports of the statements of the owners of the mines where calamities have taken place one is almost forced to the conclusion that these tragedies are the direct result of overcarefulness."[24]

Dr. Whitin and the National Committee on Prison Labor wanted the families of the dead convicts to receive compensation. Whitin asked Governor O'Neal what the proper legal steps would be. O'Neal may well have resented this interference in what he considered an Alabama affair, but he agreed to get an opinion from his attorney general.[25] Approving but wary, Whitin responded that "we trust that your investigation into the causes of the accident will result in locating the responsibility." And then, more to the point, he added:

> We are glad to know that you submitted to the Attorney-General for his opinion the question of bringing action against the contractors on behalf of the wives and children of the convicts. . . . We trust that he will feel that he can take action. . . . We may say in passing that should he find it possible to take action many people who are at present interested in the disaster and who live in all parts of the United States will feel that at least some justice has been done.[26]

The resourceful Whitin and his committee provided O'Neal with citations to cases where the courts had held lessees responsible for injuries to prisoners. And Whitin quoted from the *American and English Encyclopedia of Law:* "The lessee of convict labor is liable for the injury caused by his negligence in furnishing to the convict reasonably safe places and safe appliances in which and with which to do the work required of him."[27]

Despite the legal precedents and the obvious grounds for seeking damages against Pratt Consolidated, the state sidestepped the issue. After consulting with Attorney General Robert C. Brickell, a north Alabama lawyer from Huntsville, O'Neal concluded that "under the Constitution of Alabama the matter of bringing suits is an individual right, and the personal representatives of the convicts killed by the explosion at Banner mines would have to bring suit in their representative character."[28] The administration's sophistry was far less a conclusion from the law than it was a decision produced by policy. If the state had reason to believe that agents of the corporation had been negligent in their responsibilities, then action was warranted. The state, as the leasing authority, had every right to protect the lives of its own prisoners, and it might well have sued Pratt Consolidated for a violation of its lease contract.

There the matter stood. If litigation came, it would be the individual versus the company—the various plaintiffs would not have the state as a friend in court. The state declined to force company personnel or corporate executives to give a legal accounting for the dead. Yet two prisoners who survived did receive a benefit. John E.

Milholland, a wealthy New York Republican with a strong sense of justice for the rights of blacks, followed the details of the accident in Associated Press dispatches. He noted that Charley Brown, a black convict, was instrumental in saving the lives of numerous miners. Such heroism in the face of probable death gained Milholland's great respect. The New Yorker wrote to Joseph E. Manning—a former Alabama Populist active in reform causes, then living in Washington, D.C. He asked Manning to seek a pardon for the prisoner. Milholland noted O'Neal's good record, his Irish ancestry, and his opposition to lynching. As for Brown, "I do not know what crime this man has committed, but I do know he has atoned for every crime in the calendar that I can think of, and I sincerely hope the Governor will pardon him and pardon him promptly."[29]

Manning wrote the governor, enclosing Milholland's letter, and quick action followed. Two men, Clarence Nicholson and James Franklin, were pardoned. With irony and injustice, Charley Brown remained in prison. The paroled men had been extremely lucky to escape from the mine, but both Brown and Curlie Smith seemed to have performed the more heroic deeds. At any rate, the governor had issued an order on May 15, 1911, on Nicholson and Franklin: "For heroic service at the Banner Mine disaster, rescuing men, at the risk of their lives, I pardon these two men, conditional on their future good conduct."[30]

The prospects for the other surviving convicts, unless there was a complete change in official policy, was a return to work in the Banner. Three days after the explosion, almost routinely, eighteen new prisoners arrived at the mine. Jefferson County forwarded eleven, and the others came from three other counties.[31] No formal orders had stopped the flow of convict laborers to the mines. The possibility that Jefferson County would curtail or cancel its contracts remained only that—a possibility. Pending an official decision, prisoners were sent to the Banner.

The new men, although their normal fears were heightened by knowledge of the disaster, were less apprehensive than the survivors of the ordeal. Ed Garth, a black prisoner from the north Alabama town of Decatur, was one who dreaded a return to work. He had spent fourteen years in various Alabama prisons, serving time for petty crimes. His current incarceration marked his twelfth conviction. Garth was interviewed by a reporter who, with an inappropriate attempt at humor, still managed to capture the black's deep sense of unease. "I don't know as I would promis' to be good," Garth said, "but I believe I would, just to get free once more, for dis her coal minin' business is sho not my calling." But wasn't he an expert miner? " 'Course I can dig coal boss, but I tell you right plain sar, I sho do want to quit it, and dat real bad too."[32]

The Banner mine explosion had once again unleased the furies. It had fractured that thin but opaque shell that hid the unpleasantries from public view and that hushed the questions of reform and change. State government now stood at the center of the questions. Would the legislature end, modify, or continue the convict lease unchanged? And there was the related issue of mine safety. Statistics showed that European coal mines were much safer than those in the United States. At the same time as the Banner accident, a mine explosion at Throop, Pennsylvania, claimed seventy lives. John Mitchell, former president of the UMW, made national news with his declaration that "it is imperative that the factory and mining laws of all of our States should be enforced with the utmost vigor."[33]

The deaths at the Banner Mine gave Governor O'Neal a powerful lever. Fifty-eight years old, O'Neal was a native of Florence in the Tennessee Valley. His father, Edward Asbury O'Neal, was a brilliant lawyer who served his state well in the Civil War and was elected governor in 1882 and again in 1884. O'Neal inherited his parent's intelligence, graduated from Florence Wesleyan University, and attended the University of Mississippi and the University of Alabama. He also inherited certain physical features from his father. Full, slightly arched eyebrows and an unusually thick but well-groomed moustache set off the younger O'Neal's handsome, straight-featured face. Emmet joined his father in a successful law practice. He emerged as an excellent orator and entered politics. Loyal to the Democratic party, O'Neal was prominent in the constitutional con-

Emmet J. O'Neal, governor of Alabama, 1911–1915. (Alabama Department of Archives and History)

vention of 1901. In 1910 he was elected governor in a campaign that marked him as distinctly less progressive than Comer. Yet he and the party platform were committed to securing new mining legislation.[34]

The *Montgomery Advertiser* was the mouthpiece for the ruling Democratic party, and, as a politician, O'Neal daily read the *Advertiser.* He and the paper espoused the same philosophy, and the *Advertiser* supported his programs almost uncritically, including the plan for a mining law. If the governor needed any additional urging, he might have turned to the columns of another capital newspaper, one far less powerful than the *Advertiser:* the *Montgomery Times.* That paper editorialized, "These charred and disfigured victims of a policy passed by the parliament of hell and dictated to man by the devil, should haunt the dreams and cast a gloom over the waking hours of every Alabamian who pretends to have a conscience."[35] Although O'Neal could not command the legislators to do his bidding, he could play a forceful role if he chose. In 1911, chances were never better for enacting major changes.

4
"At the Mercy of the Earth": The Mine Safety Law

Dr. Joseph A. Holmes's appointment to direct a Washington-based agency devoted to mine safety was evidence of national concern. As one historian described the situation, "By 1910 a coalition of interests, including miners, inspectors, coal and metal mine operators, and bureaucrats, had created the United States Bureau of Mines."[1]

Alabama was particularly vulnerable to charges of legislative neglect: between 1900 and 1910, mine explosions claimed 1,180 lives in the state. In 1910 alone, 238 miners died in accidents, an average of 20 per month. Included in the grisly record were 56 killed at the Yolande mines in 1907 and another 5 in 1910; 6 deaths at the Short Creek mines in 1908 and 18 more in 1909; 41 killed at Mulga mine in 1910, a year that added 91 at the Palos mine. Another disaster occurred in 1910 when the convict prisons at the Lucile mines burned and 25 inmates perished. In reporting the Pennsylvania and Alabama explosions in 1911, the *United Mine Workers Journal* blamed the accident at Throop on failure to enforce existing laws and that at the Banner on the absence of laws.[2]

For all the protestations that the Banner was a model mine, 2 blacks and 1 white were killed on November 3, 1911, in an explosion redundantly blamed on caps and bituminite. The 128 men killed at the Banner set a new record in Alabama for a single mine accident. Of Alabama's 22,003 coal miners, 209 lost their lives in 1911, and there were 2,847 reported accidents. An additional 10 men were killed in the state's metal mines.[3] In the face of such losses, the state's mine safety laws appeared rudimentary; they were, for example, far below the safety standards of Pennsylvania.

When O'Neal accepted the nomination for governor in 1910, he spoke strongly in favor of legislation for mine safety. "Methods can be devised by which the appalling loss of life in the mines of Alabama can be lessened," the nominee said. "Such methods should be adopted regardless of cost, and mine inspectors appointed regardless of political affiliation, their selection based solely on competency and efficiency." The aspiring candidate favored stricter employers' liability laws and hinted at workmen's compensation or voluntary insurance: "Some system should be adapted by which the defenseless widows and orphans of the victims of these frequent mine disasters should be secured against penury or want."[4]

The candidate went even further in his campaign rhetoric, using words that, in fact, would soon confront him. At O'Neal's insistence, the Democratic platform included a plank that expressed his stated views: "We favor the enactment of efficient laws covering the operation of mines, which shall embody the most modern and improved mining practices, and shall adequately protect the lives of miners and the property of mine owners."[5] The need was clearly manifest; it had a progressive ring to it without verging on the radical, and few miners, owners, or politicians dissented. Production figures were testimony to the increasing industrial demands for coal. New safety problems were added to old ones. Deeper penetration into the earth meant exposure to concentrations of explosive methane gas. Increasingly sophisticated equipment expanded the amount of coal dust, which was both a health and a safety hazard. Conversion to electricity created an added danger, and emphasis on the amount of coal mined (rather than on the size of the pieces) led to the misuse of explosives.[6] O'Neal had selected the right time and the right issue.

With the governor determined to achieve results, the burden of providing the necessary laws fell on the state legislature. That body would attempt to correct part of what the *New York Herald* saw as the cause of the explosion: "inadequate laws and improper inspection and sometimes a combination of these two." The lawmakers convened at high noon on Tuesday, January 10, 1911, and would remain in session for fifty legislative days until their scheduled adjournment on Friday, April 14. In his inaugural address and message to the legislature of January 16, the governor continued to stress the theme of mine safety. Updated laws should provide for more frequent inspections, increase the number of inspectors, expand their powers, and raise their pay. To do so would attract capital and labor to the state. "Men who work in our mines," the governor declared, "should receive every protection which modern engineering skill and science can devise."[7]

Given such executive encouragement, no one was surprised when President pro tem Hugh Morrow of Jefferson County filed a bill to regulate the mining of coal. A native of Jefferson County, a graduate of the University of Alabama, and a member of a leading law firm in Birmingham, Morrow, at thirty-eight, was probably the most vigorous legislator of the session. During the term he introduced fifty-two bills and countless resolutions. Introduced on February 3, Morrow's senate bill 199 was sent to the Committee on Mining and Manufacturing, headed by Thurston H. Allen, from the Tennessee Valley city of Florence. The strong-willed Allen produced a favorable report from his committee. Read a second time on February 10, the bill was placed on the senate calendar. Then, without public explanation or comment, Morrow's bill was not heard of again until April 11, the next to last day of the session. At that time it was indefinitely postponed.[8]

Legislatures nearly always find time to deal with matters of genuine interest and influence, and the disappearance of Morrow's bill might suggest the absence of either or both. It was suggested that Morrow's time had been monopolized by other measures. Morrow was known as procorporation and was an attorney for Sloss-Sheffield Steel and Iron Company. On the basis of sponsorship alone, the bill was suspect. The *Mobile Register* explained that the measure was heavily weighted on the side of the operators, and "if it is such it is not the bill wanted by the state." It was difficult to believe that the

Hugh Morrow, state senator from Jefferson County, introduced a mine safety bill before the Banner Mine explosion. The measure died without action. (Alabama Department of Archives and History)

Morrow bill earned oblivion because it was too conservative—normally a guarantee of passage. It was more likely that the measure was killed because a more acceptable vehicle appeared—namely, the Hollis bill, introduced in the house on February 8.[9]

Representative John D. Hollis was the son and grandson of Alabama legislators. Elected from Walker County, he had worked as a teacher, as a store manager for a coal and iron company, and, more recently, as the store manager for the Pratt Consolidated Coal Company. Hollis's house bill 431 had the blessing of management and was reported to have been framed by the Coal Operators' Association of Alabama.[10] As chairman of the House Committee on Mining and Manufacturing, Hollis was one of the more strategically placed store operators in legislative history. On March 2, Hollis's committee referred his bill to the house with a favorable report, and on the same day it was read a second time and placed on the calendar.[11]

If the convicts had no say about their working conditions, organized labor spoke for all miners. The UMW opposed the Hollis bill and whatever grafts it received from Morrow's proposal. UMW officials J. L. Clemo, R. A. Statham, and J. R. Kennamer charged that the bill was forged in secret by the coal operators. Finally securing a copy of it, the union men were dismayed: "If it becomes a law we do not see how any self-respecting miner can come to Alabama." They had specific objections. The measure created a monopoly for company commissaries (often called "robbersaries" by the miners). Farmers and other noncompany personnel could not sell their produce directly to miners. A similarly protected position would be enjoyed by company physicians. No private doctors could visit mine property without company permission. The same regulation applied to candidates for office and even ministers of the gospel.[12]

The requirement that miners could not enter a gaseous mine unless safe conditions were ascertained freed the operators from legal action. Mine accidents were to be investigated, but the results could not be made public or used as evidence in court. "This law," Clemo and the others declared, "is not for the protection of the

Representative John D. Hollis, Walker County, a former employee of Pratt Consolidated, introduced the mine safety bill prepared by the Coal Operators' Association of Alabama. (Alabama Department of Archives and History)

miners, but solely for the protection of the operators against any liability for the maiming and killing of those whom they send down into . . . the earth."[13]

"If there is anything worse in Siberia, we have not read of it," the national journal of the UMW agreed. "Gotten up avowedly to protect the miners, it is a clumsy attempt to deceive the legislature and fasten a condition of slavery on the Alabama miners BY FALSE PRETENSES." In brief, the *United Mine Workers Journal* concluded that "the bill is bad, . . . and should be buried without further consideration. Kill it, and don't disgrace the state."[14]

By the time of the third reading for the Hollis bill, debate had become bitter. It was April 7, the Friday before the Banner Mine explosion. All at once a compromise seemed possible. Charles A. O'Neill of Jefferson County, representing the views of the reformers and organized labor, offered a substitute measure. His version passed by a vote of 37 to 29. Transplanted from Ohio, a first-generation American, O'Neill had been a railroad contractor, a miner, and a merchant and, at this time, operated a salvage company. Hollis, although supposedly the author of the original bill, did not vote.[15] O'Neill managed to substitute his bill for Hollis's, and he now offered amendments designed to secure its passage. His maneuvers succeeded, and on its third reading the bill passed the house by a vote of 70 to 0.[16]

While the house had voted with perfect unanimity on the Hollis-O'Neill bill, Representative William J. Martin now expressed his reservations and announced his intention to move for reconsideration on April 11. That would leave only one legislative day before adjournment, and the slightest impediment or miscalculation would kill a mine safety bill of any kind. Martin, a representative from north Alabama's Jackson County, had been born in the mining county of Shelby. He had been a teacher and lawyer in Jefferson County, and he held the office of state land agent. He was not a man to be taken lightly.[17]

The weekend that followed O'Neill's maneuvers and Martin's threat was dominated by news of the catastrophe at the Banner.

Tragedy could not have struck at a more opportune time for its effect on the question of mine safety. Governor O'Neal, well aware that the issue had been thrust into the arena of public discussion, announced that "I am going to have a mining law for the state of Alabama if I have to call a special session of the legislature. We have been delaying this important matter too long. Life is more precious than stockholders' dividends."[18] The governor turned to experts and his trusted advisers for information. One told him that "what happened is past. The future is another thing." O'Neal should make certain that the mine inspectors were adequately paid. Numerous technical suggestions came from John M. Russell, a miner for thirty-three years.[19]

A large contingent from Birmingham came to Montgomery to lobby for additions and deletions. R. A. Statham had already been at the state capital for four weeks, constantly urging the cause of labor. "We are trying to get a new mining law passed," he wrote, "and so are these union-haters, the Alabama coal operators, and we have to keep our eyes on them or they will get their bill through and kill ours, and God pity the Alabama miners if they succeed."[20]

The *Montgomery Adviser* reminded the Democrats of their platform commitment and noted that "the lives of these men, the happiness of their wives and children all point to the duty of the State to do what it can to throw more protection around them while they are at the mercy of the earth." Joining the issue, the *Birmingham Ledger* wanted the mine safety bill passed immediately: "There should be no more gas explosions and no more dusty explosions and the powder explosions can be largely protected by mine bosses and superintendents."[21]

The careful negotiations of O'Neill, the forceful statement by the governor, and the editorial support from the press seemed to pay off. After the senate received the bill on Friday, it was referred to Allen's committee on Mining and Manufacturing. On the following Tuesday the committee gave the bill a favorable report. It was read a second time and placed on the calendar.[22]

On Friday, the final day of the session, the last act in the labyrinthine legislation took place. Allen's committee now offered a substitute bill, a clear signal that pressure and negotiation had continued without interruption. There were not many changes in the Hollis-O'Neill bill. The number of mine inspectors was increased, although whether that favored management or labor remained to be seen. One clause that specifically held coal mine owners liable for damages was taken out, which represented a major victory for the coal companies. The committee bill was passed by the senate without dissent.[23]

When the bill was returned to the house for final action, Representative Martin's threat of opposition never materialized. The changes made by the senate (the bill now incorporated language from the dormant Morrow bill) and perhaps the public outcry over the Banner had placed Martin in a more cooperative mood. Representative Hollis, who had abstained from voting on the bill's amended versions, now championed the measure. Once the governor signed it, Hollis predicted, "Alabama will have the best mining laws in the United States." The only negative vote came from Representative Simeon T. Wright, a lawyer from Fayette, as the bill passed the house 67 to 1.[24] Alabama's new mining law, most observers agreed, was an amalgam of the views of the governor, the mine owners, and the miners. The generalization was correct, but it said little and explained less. The real question concerned the relative proportion of those views.[25]

Later, O'Neal would take full credit. He said that he had called together representatives of labor and management. "I secured copies of the most advanced laws in the union on this subject, as well as the statutes of foreign lands, and after a conference, lasting over a day and night, succeeded in drafting a bill which both the representatives of the operators and of the coal miners endorsed." Not all labor leaders endorsed the compromise. R. A. Statham opposed the bill in its entirety and excoriated Edward Flynn, former president of UMW District 20 but now an employee of TCI, for supporting it.

Representative O'Neill took a realistic stance: "I had hoped to have a bill passed that would satisfy both the miners and the operators, but one cannot get everything he wants."[26]

A pleased Governor O'Neal signed the bill into law on Tuesday, April 18. There was widespread approval across the state. A Birmingham paper demanded "strict enforcement, no matter what the cost," and W. R. Fairley, a former member of the UMW Executive Board, told the Alabama Federation of Labor that mine safety would be much improved. Fairley's approval was seconded in a speech by J. R. Kennamer, president of District 20. Neither of the men was enthusiastic, but each was practical. The law was not perfect and it was long overdue, but, as the *Talladega Reporter* concluded, better now than never.[27]

Given the power of the owners, how had a bill passed that would increase their operating expenses and reduce their profit margins? Governor O'Neal exaggerated when he wrote that the "prompt passage of this law was secured by my statement to the coal operators that unless it was enacted, I would immediately convene the legislature in extraordinary session."[28] Had the operators genuinely desired to defeat the bill, they could have girded themselves for a fight and might well have won. Yet, in a sense, O'Neal was right. An extra session would have brought an unusual focus on mine safety, and to oppose it would have been bad public relations.

Inevitably, the related subject of convict leasing would have come up. The potential debate over drastic change in the labor system was something to be avoided. It was better to let that issue receive the divided attention of the legislators and the citizenry amid the confusion of the regular session. Also, in the progressive spirit of the age, mine safety was a tangible subject, one that could be comprehended and dramatized. The law was an act of political pragmatism. Spokesmen for the miners pointed out that, without the act, future mine disasters "can be laid at the door of the legislature and the administration." Nor could it be denied that most of the operators, including McCormack and Ramsay, were committed to mine safety. Beyond that, the owners were not crippled by the law's provisions.

As Representative O'Neill candidly said, "There is considerable power given to the inspectors, but the operators are not to be run over by any means."[29] Finally, though, had there been no tragedy at the Banner, there would have been no mine safety law in 1911.

The new law seemed comprehensive. It increased the number of mine inspectors from three to seven (a chief and six assistants). The inspectors were to serve for three years with staggered appointments, and the positions were considered to be full-time jobs. The chief inspector's salary was set at three thousand dollars annually, with the associates paid two thousand. Every mine was to be inspected at least every three months. Strict regulations were spelled out for every phase of coal mining: qualifications and requirements for inspectors, reports, investigations, foremen, fire bosses, operation of machines, shafts, props and timbers, openings, ventilation, fencing, doors, furnaces, pipes, airways, safety lamps, cages, brakes, chains, tubes, cables, maps and surveys, explosives, fans, and scales. Allowing the chief inspector (with the agreement of two associates) to halt a mine's operation if too much gas or dust accumulated was viewed as a concession to the miners. No woman of any age and no male under fourteen years of age could be employed in a mine.[30]

Representatives of the miners' viewpoint were able to delete some of the blatantly promanagement clauses of the act's earlier versions. They were not totally successful, however, as evidenced by section 11. This section prohibited the release of any inspector's report of a mine accident to the public. Such information could be revealed only in a legal proceeding; except for next of kin being notified, there would be no publicity. Several mine accidents that resulted in fatalities occurred shortly after the law went into effect, and newspapers were unable to obtain details. One editor angrily wrote, "The people don't know enough to be knowing of accidents happening in mines; in their ignorance they might become prejudiced against some of the benevolent methods of the Steel Trust and other corporations; therefore all knowledge of the slaughter going on must be kept from them."[31]

The most glaring absence from the law was the provision dropped during the senate compromises that would have established clear-cut liability for the coal companies. Governor O'Neal's earlier heroics for a system "by which defenseless widows and orphans of the victims . . . should be secure against penury or want" had been effectively muted.

The new mine safety act occupied forty-three pages in the official *Acts* of the state legislature for 1911. Throughout those pages the word "miner" was used frequently but always in the abstract sense. It described a form of employment rather than a laborer's personal condition. It did not differentiate between the free worker and the leased convict. Yet that difference remained as a crucial question, a matter of concern for many both in and out of Alabama. The battle over the convict lease system was fought in the legislature simultaneously with the issue of mine safety. It had a different outcome.

5
"A Principle of Justice": The Convict Lease Question

Alabama was cursed by peonage as well as by the convict lease system. Peonage, a self-induced trauma, also had its origins in the exploitation of human beings for profit. Complex and confusing, peonage rested on debt and assumed different shapes. A sharecropper unable to pay his debts could be forced to remain on the owner's plantation. Local jails were the seedbeds for another form of peonage. Inmates permitted prospective employers to pay their fines; in return, they obligated themselves to work out the amount paid. The potential for exploitation was limitless and made for scandal and exposé in the twentieth century.

Whereas peonage operated in the shadows, convict leasing was boldly paraded, and Alabamians had less excuse for tolerating it. Nor could its defenders in Alabama seek the comfort of numbers. One by one, other southern states had eliminated it or were in the process of ridding themselves of the system. Tennessee abolished leasing in 1893, largely as the result of armed uprising by free coal miners. Louisiana's constitution of 1898 required the system's termination in 1901. The hiring out of prisoners was dropped in Georgia in 1909 and the next year in Texas. Arkansas would make the move in 1913.[1]

Even as neighboring states abandoned the convict lease system, Alabama clung to it publicly in the name of fiscal responsibility and from the private motives of entrenched interest and personal greed. It was a perennial political issue—as new as each meeting of the state legislature, as old as the horrors of early railroad building and the hypocrisies of Bass and Bankhead. In the session of 1911, long before the tragedy at the Banner, two bills on the matter had been introduced. One sought to abolish the lease system; the other hoped only to cure its minor injustices and to rationalize its ambiguities.

Neither piece of proposed legislation was the result of executive leadership. Governor Braxton B. Comer, in his farewell address to the legislature on January 11, 1911, had dismissed the "complaint that the convicts should not be worked in the mines because of competition with the labor in the mineral sections." The truth, according to Comer, was that "the state convicts are well cared for and their labor tasks are not unreasonable." The money derived from convict leasing was "used in the general expense account of the State and helps maintain the schools, pensioners and judiciary." Using an argument of ancient lineage, Comer warned, "Should this fund be cut off or diverted, its place would have to be supplied from some other source or the interests of the State . . . will suffer."[2] As for the incoming Governor O'Neal, he treated the subject in time-honored manner. He did not mention it in his message.

It was not from executive leadership or support that Senator Robert Moulthrop, drawing on his own conscience, introduced senate bill 56 on January 18. Moulthrop was a native of Georgia, college educated, and represented Barbour County, a largely agricultural but politically sophisticated region in the state's southeastern section. He was a prosperous brick manufacturer and a civic leader in Eufaula. Moulthrop's bill called for a sharp break with the past: prisoners convicted of felonies and misdemeanors (excluding certain classes) were to be employed on the public roads of the various counties; the hiring of convicts to private persons or to corporations was specifically prohibited.[3]

It has been suggested that Moulthrop's purpose was more pragmatic than humanitarian, that he specifically wanted to provide adequate farm-to-market roads for his rural constituents. In fact, if rural road building had ever been a significant interest, the convict lease system might have been abolished decades earlier. There is no

Senator Robert Moulthrop of Barbour County introduced a bill and led an unsuccessful fight to end convict leasing in Alabama. (Carnegie Public Library, Eufaula, Alabama)

need to deny Moulthrop his strong sense of justice in favor of his economic impulses.

Moulthrop saw his bill as a cause: it "appeals to humanity [and] seeks to give justice to the unfortunate." "I believe," said Moulthrop, "in respecting the legitimate rights of free labor. I do not believe that the State of Alabama has a right to compel the honest free laborer to either compete with a convicted man or else seek new lines of employment." Moreover, "I don't think the state has a right to fill its coffers with the money earned by these convicts. It savors of blood money. . . . I do not think it is right for a convict to be forced to work under conditions in our coal mines which free laborers will not yield to. There is a principle of justice involved and neither side should be made to suffer unnecessarily."[4]

Once introduced, Moulthrop's bill entered the Byzantine channels of the Alabama legislature. It might logically have been referred to any one of three committees: Mining and Manufacturing; Penitentiary, Prisons, and Punishment; or even Public Roads and Highways, of which Moulthrop was a member. Instead, with unexplained logic, it was sent to the Military Committee. (The reason could scarcely have been that armed guards oversaw convicts; more likely, the reason lay in Moulthrop's membership on that committee as well.) The committee was chaired by William J. Vaiden, a Spanish-American War veteran and an aristocratic planter from Uniontown in the Black Belt county of Perry.[5]

The referral of bills to inappropriate committees has been a standard ploy of legislative leaders seeking to lose a bill or to kill it completely. In this case, the strategy did not work. On February 1, Vaiden's committee reported Moulthrop's bill favorably. It was read a second time and placed on the senate calendar. To that point Moulthrop's volatile piece of legislation had enjoyed a remarkably easy progression. On February 3, its inexplicable success story began to change. On that day, the senate responded to a motion by John J. Espy of Headland (Henry County) and recommitted the bill to committee. It was this very plasticity of action that made legislative moves so confusing and misleading. Although passage by committee was not tantamount to senate approval (nor did recommittal mean death), endless action and counteraction allowed legislators to mask the meaning of their votes. Interest-group motives were thus buried in a thicket of words—support and endless rejection.[6]

The immediate fate of Moulthrop's bill now rested in the hands of Senator Charles M. Sherrod, the chairman of the powerful Committee on Revision of Laws. Sherrod was from Courtland in Lawrence County. The long-time lawyer-editor-politician presided over the committee that decided on the future of recommitted bills.[7] Bill 56 received reconsideration, and on February 8, the committee ordered it returned to the senate with a favorable report. Moulthrop was not satisfied with what appeared to be his bill's miraculous escape from death, and he attempted to speed up the proceedings. On February 21, he was able to enter a motion under a suspension of the rules to consider his bill on the following day. In fact, it was two days before the senate considered it, but on February 23, the bill was read a third time and passed the thirty-two member senate by a vote of 12 to 9.[8]

The senate of the state of Alabama had now passed a bill abolishing the convict lease system. The reform of the decade seemed at hand. Yet passage aroused neither approbation nor opposition. The press did not launch a blizzard of editorials or hang expectantly on the outcome. Alabamians knew that it was the last vote that counted. The decision of a truncated senate to do what no one thought possible was only a matter of passing interest. The sequel proved how right they were.

On March 1, the house received senate bill 56 and sent it to its Committee on Public Roads and Highways. Chairman George Washington Darden, a lawyer and politician from Oneonta in Blount County, and his committee approved the bill on March 10 and returned it to the house.[9] And there the action stopped. The journal of the house contains no further mention of the measure, no second or third reading, no recommittal, and no voting results. The inference is overwhelming: the bill was killed by inaction, probably the plan of house and senate leaders all along. Open opposition had always been the crudest form of action—and the least necessary. Moulthrop was two decades ahead of his time.

Where Senator Moulthrop sought the complete abolishment of the lease system, Representative William J. Martin of Jackson County sought improvements in the existing structure. Martin was hardly a leader among Alabama's reformers and progressives, as his lack of enthusiasm for the Hollis-O'Neill bill testified. He was, however, the chairman of the house Committee on Penitentiary and Criminal Administration, and on January 16, only two days before Moulthrop, he introduced an important bill.

What can be known of the Martin bill was contained in the house journal and the résumé of the bill it provided. There was no question that it specifically provided "for a continuation of the present method of employing, supervising, guarding, hiring out and working the State convicts." The different thrust of the bill was that all costs of conviction (except rewards), including the continuing expenses of imprisonment, would be paid to the state and to the county where the conviction took place. The probate court would then take 50 percent of the convict's net earnings and pay the money to the prisoner's family or dependents. If there were no dependents, the money would go into the common school fund in the convict's home county.[10]

The measure could be interpreted as an effort by the state to gain funds that had previously gone exclusively to the counties, for it left a major point unanswered. If court costs were coming off the top of the convict's earnings and 50 percent of the net went to families, where did the other 50 percent go? If the money went to general county funds rather than paying off additional court costs, the results were predictable. Sentences would be grossly prolonged because of the time extension necessary to repay costs. Whatever the ambiguities, two things were clear about the Martin bill: for the first time it would have provided for a convict's dependents, and it obviously constituted a major attack on where the money went.

Martin's bill was sent to his committee, and on February 9, it was reported to the house. A month later, on March 7, the house took up the bill, and it was quickly apparent that the proposal was no approved party measure.[11] Representative Boswell deGraffenreid Waddell, a lawyer from Seale in Russell County, moved to postpone consideration until the next to the last working day of the session (April 11)—a standard ploy to kill legislation in the rush and confusion of adjournment. Martin managed to get Waddell's motion laid on the table, and the noon recess left his bill still under debate.[12]

When the house reconvened that afternoon, the bill weathered a stylistic amendment, but its real difficulties began when Representative John V. Smith stood and was recognized. Smith was one of the most influential men in the house. Born in Georgia, Smith was a lawyer, a former president of the Alabama Railroad Commission, and a former president of the State Board of Pardons. From his power base in Montgomery, Smith was a key activist in the inner councils of the state Democratic party. He now moved that house bill 70 be laid on the table. The house concurred, and at that point, on March 7, Martin's measure disappeared forever into legislative oblivion.[13]

The bill's fate was clear evidence of the impenetrable nature of the Alabama legislature when major interests were at stake. Martin was a conservative and not at all imbued with crusading zeal. His bill had made no change in the operation of the system. Yet what Martin had done was to attack the very vitals of the institution—he wanted to change where the money went. From the recipients' view, that was the most important matter of them all. Whatever the state did with its prisoners was its own business, but county convicts should be dealt with by the counties. The merest thought of Martin's changes must have caused fear and trembling (and steely resolve) in every courthouse ring in the state.

It was obvious that reform sentiment on the convict lease system was hardly a match for the deeply entrenched status quo. The issues of society often seem to rest on a seesaw, now heavily weighted at one end and skewed in that direction, now slowly rising as some added pressure moves the issue into equilibrium or on to reversal and change. In early 1911, there were no weights causing the side of reform to rise. The newspapers and the public pressure that they fueled were silent. Crusades are episodic; the attention span for reform is incredibly short as new sensations rise to replace the old.

With the tilt toward inaction, the explosion that ripped the Banner Mine threw the heavy weight of shock and surprise on the side of change. Disaster created an air of immediacy. Reopening the question, a Republican newspaper repeated Moulthrop's arguments and called for a change: "It should not be part of the sentence to put inexperienced men in places of danger. The State cannot afford to take the risk of having its convicts killed. Nor is it a square deal to the experienced miners to be forced into competition with convicts." H. M. Wilson, editor of the *Opelika Times* and a member of the convict board, believed that the sentiment of Alabamians "is overwhelming against the working of convicts in mines." Another editor wrote that "the spirit of commercialism has eaten too far into the system of dealing with our criminal classes."[14] The voices of change were in the air.

In the days immediately after the Banner disaster, Representative William Barnard of Jefferson County moved to translate the outcry of the state press into legislative action. The twenty-seven-year-old Barnard had run a livery stable, had served as an undertaker, and had operated as a coal merchant. If additional toughening was required, he had also served as a deputy sheriff and chief of police.[15]

On the next to the last day of the session, Tuesday, April 11, Barnard introduced house resolution 128. It called for an end to the convict lease system in words that might have been written by the executive committee of the UMW. Barnard charged that the loss of lives at the Banner was directly attributable to the lease system. To place state and county prisoners in competition with free laborers was "an unwise policy." If the legislature adjourned before action could be taken, the resolution asked the governor to call a special session. Barnard's proposal was quickly referred to the Committee on Rules, and there was some support in the state for a special session.[16]

Even so, invoking the arguments of justice, compassion, and every instinct of humanitarianism was plainly not enough. Echoes of former governor Comer's speech still reverberated within the capitol walls: if the convict lease system was abolished, what would take its place? To confine state and county prisoners to a penitentiary would eliminate the lucrative revenues that leasing provided and would place a continuous burden on the state treasury. The progressives were advocates of putting the prisoners to work on public roads and highways—an answer that satisfied their social consciousness with practical results. The state would receive the full benefit of their labor, and—a most naive thought—the convicts would be better protected because their treatment could be scrutinized by the general public.[17] The lines were drawn between the supporters of convict leasing (and their allies, the doubtful and the undecided) and the reformers. The question was whether the balance would change—or could be changed.

Avoiding the risk of alienating anyone, some newspapers took no stand at all. The *Northport West Alabama Breeze* championed the less than bold proposition that mine disasters ought to be prevented. The *Selma Journal* presented a classic case of ambivalence. The paper admitted that to consider income over other factors was wrong. That would be sordid and "putting a money price on the more precious personal right—the inherent right to live." But was road construction an adequate alternative for leasing? "When the convicts are placed on the public roads they are going to become an expense to the state instead of a source of income." Still, "the income . . . will be great through saving what is expended on roads, but this would be indirect. In other words the income now is direct. Whereas when the change is made the expense will be a direct outlay of money, and the return therefore will be indirect." Readers who managed to follow the tortured reasoning to that point discovered that the Selma paper did not know what to do or what to suggest. It had argued itself into equilibrium and retreated to blind optimism: "We hope some plan can be speedily evolved by which the lives of the convicts can be guarded in the future."[18]

Widely circulated urban newspapers were among the unblushing defenders of leasing. The *Montgomery Journal* admitted that many of the convicts had committed minor crimes and acknowledged that some were innocent, but it questioned any policy that reflected too much sympathy. "Must they be treated any better than the free laborer?" After all, "A large number of them are hardened criminals." That rationale would have drawn rebuttal from the families of four blacks from Butler County (George Gilmore, George Broughton, Garfield Smith, and Will Sims) and of one white (John Howard). Each man had involuntarily exchanged a one-year sentence for eternity.[19]

Yet the *Journal's* logic drew acceptance from its influential city rival, the *Montgomery Advertiser.* That paper refused to follow "a sentiment which demands that criminals should be coddled at the expense of honest men. . . . They are prisoners because they committed crimes, they are where they are because crime should be punished. . . . They have a right to protection, but they have no right to petting and coddling."[20] Until a practical substitute could be devised (ideally one that did not pet and coddle), the *Advertiser* favored sticking with the lease system.

Putting prisoners on the public roads was "no practical substitute" for the *Birmingham Ledger.* This solution would be costly and would have undesirable political ramifications. The *Ledger* reviewed Jefferson County's brief experiment with using prisoners on the public roads. Seeing men dressed in striped uniforms had been demoralizing to the public, and they were soon removed. There was no worthwhile alternative to the lease system. "Men who commit crimes should pay the cost of their own punishment. Law-abiding people feel sympathy for convicts, but few of them are willing to be taxed to support the criminals."[21]

The ultimate recipients of the monetary bonanza of convict leasing, the mine owners, had not usually rushed to defend the system publicly. In an earlier day John T. Milner had entered the lists in defense and praise of using prisoners to mine coal, but most of the entrepreneurs had counted their profits and left their political allies to handle affairs in the public arena. Now, in 1911, John Campbell

Maben, Sr., staunchly backed the lease system. As befitted the chief executive of Sloss-Sheffield Steel and Iron Company, one of the district's largest corporations, Maben was forceful and direct.

On first coming to Birmingham, the native Virginian and Princeton alumnus explained, he had opposed the use of convicts in the mines. Now he favored the practice. Convicts at Flat Top, a Sloss-Sheffield mine, were paid and could leave with money in their pockets. Once released, the former convicts could turn their training into jobs on the free labor market. The corporate executive's statement was correct. A significant proportion of Jefferson County's free labor force consisted of convicts who had served their time and remained in the area to work. Continuing, Maben insisted that working in a coal mine was not a perilous occupation. Once a work shift was completed, the convict was free the rest of the day. Maben painted an idyllic existence, suggesting that crime might increase as men hurried to take advantage of being leased to the mines. In fact, he strained credulity with his inference that prison labor was a desirable form of apprenticeship and that coal mining did not threaten life. Maben considered the convicts' living conditions—"they have spacious halls to frolic in"—and concluded that "they are well satisfied as a general thing."[22]

The foes of the convict lease system were aggressive and articulate. "The convict law in Alabama is a ferocious animal constantly seeking new victims," said the *Mobile Register*. That south Alabama newspaper had long been a leader in the fight to end convict leasing. Deploring the deaths at the Banner, the *Register* would accept no rationalizations: "There is no excuse that the event was unforeseen. . . . This is official murder." Perhaps the Banner tragedy would bring needed action. "It is a sad commentary on official life . . . that such a horror is required to move impulses which long should have been set in motion by instincts of humanity, but now that better impulses have been moved and the state awakened to its duty by shame, it is to be hoped that no time will be lost in removing the blot of the convict system from the state of Alabama."[23] The *Register's* sentiments were repeated in different form across the state.

The reformers who demanded change granted the argument that the Banner was one of the state's safest mines, but they turned it to their advantage. If the mine itself was safe, then, the *Tuscaloosa News* deduced, "The blame can be put only on the system of leasing the convicts to work in the mines." What would be the fate of mines with fewer safety devices and alarms? To one editor the answer was plain that "the State is doubly guilty if it does not take the convicts from the mines at once." Another editor declared that the state dollars received from leasing "are bedimmed with the tears of widows and orphans and stained with the blood of the oppressed"; only by removing the prisoners from the mines could there be a "safeguard against repetition of this bloodletting."[24]

Some business leaders in Birmingham, such as G. M. Hudson and Mims B. Stone, saw advantages in employing the prisoners to build good roads—"something that will be of benefit to the people," in Hudson's words. Hudson and Stone and others with similar views drew praise from an unexpected ally: organized labor. As expected, the UMW was a strong proponent of terminating the leasing of all convicts. R. A. Statham and Henry Bousfield wrote powerfully reasoned letters to the *Birmingham Labor Advocate,* and the journal added its own editorials supporting passage of the Moulthrop bill. As senator from the rural county of Barbour, Moulthrop found to his surprise that he had become a hero of the industrial proletariat.[25]

One possible solution was to copy the intent, if not the actual practice, of Mississippi. That state's constitution of 1890 became famous for its elaborate legal means to disfranchise black voters, but it also set the year of 1894 as the termination of convict leasing, a practice dating from Reconstruction. Mississippi's prisoners had been leased and subleased to plantations and railroad contractors. Supposedly, after 1894, leasing was illegal, but the prohibition was violated as state officials leased lands from individuals and permitted them to manage convict labor and share the profits. An investigation ensued in 1902, and under the leadership of Governor James K. Vardaman, the practice was stopped in 1906. A series of state agricultural units, such as Parchman Farm, was established. The system provided financial profits and supposedly trained the inmates in scientific methods of farming.[26] Such an operation, building on the Williams's farms of the past, seemed viable for Alabama. Any substitute was preferable to convict leasing, according to a citizen of Talladega who insisted, "This is one reform that must come." Although nothing could restore the lost lives of the prisoners, *Howle's Iconoclast* claimed that their blood called "for the safety of those that follow them. It calls for the removal of the convicts from the mines where their lives are in constant danger."[27]

During most of the 1911 session, the legislature, subjected to no pressure from governor or public, had done nothing to change the system. The efforts of Senator Moulthrop and Representative Martin to abolish leasing or to ameliorate its effects were both defeated. The issue of mine safety had been ignored as the Morrow bill floundered in the senate, and it had taken the explosion at the Banner to force the passage of the Hollis-O'Neill bill and give Alabama an effective mine safety law. Many reformers hoped the Banner tragedy would exert enough pressure to gain action on leasing as well. When the legislature adjourned on Tuesday, April 11, there remained only one session, Friday, before adjournment. The only positive action was Representative Barnard's resolution calling on the governor to convene a special session, but this resolution had not been voted on. The statewide reaction made it clear that, while convict leasing had its strong opponents, it still had powerful support. The politicians now had to decide whether the time had come for change.

In common with legislative bodies then and now, the Alabama House of Representatives had far more business pending than one day could accommodate. Several measures would require extended debate. To cope with a crowded calendar, Representative William O. Mulkey, from the wiregrass bastion of Geneva County, advanced a plan for action. House members could caucus during their free days of Wednesday and Thursday and set their legislative priorities. For pending bills on which there was general agreement, debate would be limited, leaving time for discussion on controversial matters. With luck, the caucus could provide for an expeditious and harmonious final session.

Some members of the house were dubious about the caucus idea, but a majority agreed to the plans. The morning session on Wednesday was poorly attended. No more than forty representatives made an appearance, and several declared that they would not be bound by any decisions. One of Governor O'Neal's supporters, Alexander Pitts of Dallas County, served as chairman. He was aided by Speaker of the House Edward B. Almon, another administration man. The afternoon session was even more disappointing, as no more than twenty-five legislators appeared. There was a much larger turnout on Thursday. The caucus agreed on a few matters, but the big argument, lasting for two hours, was on convict leasing.

There was some interest in the caucus for a special session, but this was eclipsed by the possibility of recalling the Moulthrop bill from limbo. That would avoid a special session and would effectively abolish the convict lease system. Judge S. Williams, a lawyer from Clayton in Senator Moulthrop's home county of Barbour, led the fight for the Moulthrop bill, with William O. Mulkey, the father of the caucus, heading the opposition.[28]

Leading off for the conservatives, Representative Mulkey argued that the Moulthrop bill was far too radical, although he grudgingly admitted that the convicts might at least be removed from mining. He was opposed to working them on the roads. Chairman Pitts was willing to support a legislative resolution instructing the state Convict Board to stop leasing prisoners to the mines. But because Moulthrop's bill would take money from state schools, Pitts objected to any action that "would favor thieves and murderers against the children of Alabama." Frank Stollenwerck, Jr., a Montgomery County lawyer, banker, and lumber executive, also opposed the public roads alternative. Stollenwerck believed that placing prisoners in turpentine camps would yield the state as much money as sending them to the mines.

Representative Peter B. Mastin of Montgomery County saw nothing to recommend the Moulthrop bill: "I have no tolerance for it. It will bankrupt any county on earth to guard and protect the convicts if they are put on the public roads. This silly sentiment I think is largely caused by a lot of silly women." Had Julia Tutwiler been present she would have smiled. Representative Fleetwood Rice of Tuscaloosa announced that labor in coal mines was no worse than labor in cotton mills. As Rice saw it, "Thousands of good men work in mines, and I can see no reason why murderers should not do so." There was nothing unjust in the assignment. "The convicts are lawbreakers and we send them to work out their sentences. There are far more accidents on the face of the earth than in the mines. This is all sentiment."

Representative John W. Green, a Dallas County locomotive engineer, argued that there were positive virtues in leasing convicts to mines. He enumerated the benefits that caused inmates to prefer coal mine work to that on highways: sanitation, access to books, and the availability of religious services were superior in mining camps. Representative Thomas H. Molton from Jefferson County then suggested that prisoners working on the roads constituted a major threat to the public: "There is constant danger of the convicts escaping."

The supporters of the Moulthrop bill also had their say. Joseph H. James, Jr., of Uniontown, whose distinguished genealogical credentials gave the intimation of conservatism, attacked the lease system as human slavery. John H. Cranford, a Walker County merchant and banker, and president of a coal company, rejected Boswell deGraffenreid Waddell's "loss of state revenue" contention. He issued the stern reminder that human life could not be measured in dollars and cents. George W. Darden, whose Public Roads and Highways Committee had favorably reported the Moulthrop bill, added that convict lease revenue was "blood money." Prisoners should be put to work on the public roads of Alabama. James E. Jenkins of Bullock County, one of the few Civil War veterans in the legislature, believed that the Banner disaster was providential: "It was caused by a just God to arouse the conscience of this legislature."[29]

The strategy of the reformers was to persuade the caucus to accept the Moulthrop bill and to place it on the house agenda with other measures requiring only limited debate. In the aftermath, some observers believed that this would have succeeded if it had not been for Josiah J. Arnold, an administration supporter from Calhoun County. Arnold argued persuasively that small counties could never stand the financial strain of hiring guards for road gangs. He gained more support with his sentiment that it was unwise to bring such a serious issue before a dying legislature. On that solemn thought Arnold quickly moved for an adjournment sine die. His motion passed, and the caucus broke up. It seemed a foregone conclusion that debate on leasing would resume when the house met on Friday. With only one day for argument, a reporter's prediction that there would be "a long and bitter fight" was hardly overdrawn.[30]

As matters turned out, the final day of the session was more festive than serious. The old capitol, destined for renovations during O'Neal's administration, was filled with visitors. Stylish women dressed in their spring finery crowded the galleries. Presents were distributed—pages and clerks presented Speaker Almon with a gold-headed cane and a gold necktie pin. President pro tem Hugh Morrow accepted a handsome scarf set, and ex officio senate president Walter D. Seed received a masonic ring. A booming business was carried on by the Baptist women of Montgomery from their lunchstand in the rotunda. Later, when the legislature went into mock session, Representative Martin entertained the audience by playing a fiddle and blowing his fox bugle.[31]

In the midst of celebration, the lawmakers passed some bills to send to the governor. Moulthrop's bill was not among them. The supporters of the measure, led by Representative Williams, attempted to bring the bill to the floor for debate. Their motion was defeated on a voice vote, and Speaker Almon rapped his gavel for the next order of business. When the legislature adjourned that night, convict leasing remained unscathed and unaltered. Once again efforts at reform had been shrugged off by the redoubtable system.[32]

When debate finally trailed off and the last speech had been made, it was the economic argument that prevailed. To cancel convict leasing was to cancel a guaranteed source of income. Governor O'Neal did nothing to persuade the lawmakers to eliminate that source. For him, the mine safety law was more than sufficient. With it in place, O'Neal wrote, "The lives of the convicts in our mines will be properly safe-guarded."[33]

As a realist, Senator Moulthrop understood the potency of money and influence—and fear. When he had time to ponder the reasons for his bill's defeat, Moulthrop did not spare the mine owners. The operators were respected men. Declaring that only leased convicts stood between unending strikes by organized labor and industrial peace and profits, the owners convinced the legislature that they were right. Moulthrop said that "the corporate interests do not love the dear people of Alabama with that devotion that they are willing to pay more for convict labor than they can obtain free labor for. They hold the convicts a regulator over the just demands of free labor, and the state of Alabama is particeps criminis [an accomplice]."[34]

Moulthrop was not so detached on the session's last day as he sat helpless before the rules of parliamentary procedure. In an interview he reacted with both bitterness and eloquence. Convict leasing was a discredited and unjust system, and "the question cannot be choked down," he remarked. "My measure was finally smothered to death in the House this afternoon by the adoption of trickery and politician machination. The issue, though temporarily down, is the one living, vital question before the people of Alabama."[35]

6

"According to the Best . . . Experts"

As of April 18, 1911, Alabama had a new mine safety law. Published by the state as a forty-eight-page booklet, the act was indexed and annotated at no expense by the Alabama Coal Operators Association. That fact was prudently unpublicized, and the original printing of one thousand copies was raised another thousand to supply the state's 325 mines.[1] For all its flaws, the measure was the best the legislature had ever enacted. The impassioned and acrimonious effort to end convict leasing had failed, but the Banner Mine tragedy brought into question whether Pratt Consolidated would be permitted to continue hiring prisoners. If official investigations, as opposed to private speculations (informed and otherwise), placed the blame on the corporation, then, presumably, it would be denied the use of convict labor. The cause of death for the 128 men could not be left unresolved, and the search for answers went its tortured and bureaucratic way.

It has been noted that President McCormack quickly issued an official statement. Within two days after the explosion, he publicized a defense of Pratt Consolidated: the accident was caused by individual negligence. If the specific individual or what his act was or when and where he committed it could not be pinpointed, McCormack at least exonerated the company from any guilt.[2]

Erskine Ramsay, who had considered the Banner his personal showcase, did not attempt to explain how the accident occurred or who was responsible. Instead, he wrote a lengthy, scientific article on how mines were operated. Besides dealing with technical matters, the vice president and chief engineer of Pratt Consolidated explained the beginnings of the Banner Mine in 1904 and the increase of its productive capacity.[3] But it was McCormack who spoke for the company.

As previously noted, Pratt Consolidated hired convicts from counties across the state but depended on Jefferson County for its major supply. The future and unwilling miners were men sentenced in the county court, court of common pleas, justice of the peace courts, and the city court of Bessemer. Those companies that leased prisoners made quarterly payments to W. D. Morrow, a county official whose title exactly matched his job: hard labor agent. In turn, Morrow turned over the funds collected to the Jefferson County Board of Revenue. Occasionally, small payments were made for a single prisoner or perhaps one or two, but usually it was a lump sum representing a large number of prisoners. In the quarter ending just before the Banner Mine accident, for example, Pratt Consolidated paid Hard Labor Agent Morrow $5,184.90.[4]

No coal companies obtained convicts from Birmingham. Sensitive to its reputation as the New South's "Magic City" and facing constant opposition from organized labor and image-conscious civic leaders, the city did not lease its prisoners. Merchants reasoned correctly that free miners with money in their pockets meant more to the local economy than convicts. City officials abandoned tentative experiments with leasing and finally settled on employing prisoners on the city street gang. The Board of Revenue, which replaced the Board of Commissioners, was the major governing agency for the county. The board made leasing contracts with several coal mines in the area, including Pratt Consolidated, TCI, and Sloss-Sheffield Steel and Iron Company. Rather than put its convicts on the county roads, the Board of Revenue believed that leasing the prisoners was too important a source of income to give up.[5]

Job Going, president of the board, announced that his group would discuss the issue of continuing the lease with Pratt Consolidated. He promised that the company's report would be considered. Even with that proviso, Going doubted whether it would be possible to break

Convicted in city court, these prisoners worked on the streets of Birmingham, not in the coal mines. (Birmingham Public Library Archives)

the contract and speculated that the convicts would probably return to the mine. Almost idly, Going wondered if the county would recover full pay for the convicts who were killed. Perhaps the contract would be modified. The county attorney, J. H. Miller, was looking into that angle. Some members of the board accompanied B. L. Brasher, coroner for Jefferson County, out to the Banner Mine for a preliminary inspection. While the decision was pending, prisoners continued to arrive at the Banner, and the general mood was that the Board of Revenue would not (perhaps could not) break the contract.[6]

Coroner Brasher took center stage as the Board of Revenue waited for others to investigate and draw conclusions. Brasher announced that in conformity with existing laws, his coroner's jury of six men would visit the mine on Saturday, April 15, to determine the cause of the deaths. The jury consisted of Foreman J. G. Moore, Thomas Worthington, W. B. Leake, Milton H. Fies, D. A. Echols, and F. M. Jackson. Predicting an all-day session with twenty to thirty witnesses, Brasher hoped that the verdict "will be more intelligent than at former disasters."[7] Because several members of the Board of Revenue had already made personal inspections, only one accepted Brasher's invitation to accompany the coroner's jury.

On the appointed day the investigators came into session at the Banner with the inquest specifically directed to the cause of Ernest Knight's death. The jury members entered the mine, examined the damage (most repairs had already been made), and heard from a score of witnesses. With evidence in hand, the jury issued its verdict. Ernest Knight had died as the result of an explosion of blasting powder in seven left.

The jury found that, prior to the explosion, the mine had been in good condition. The mining laws of the state were being carried out. Specifically, the jury concluded: "We have considered that the mine was in good condition and that the Pratt Consolidated Coal Company was using every reasonable means for the prevention of accidents and was operating this mine according to the laws of the State of Alabama and the instructions of the mine inspector."[8] It followed

The tipple at the Banner Mine. (Birmingham Public Library Archives)

from the jury's verdict that the company was not at fault. The cause of the explosion lay with the ineptness and malpractice of the workers.

Few had challenged the defense of the company made by President McCormack, and in the public discussions of cause and blame, the weight of opinion exonerating the mine owners was becoming overwhelming. If the Board of Revenue had been ambivalent, its resolve was strengthened by the coroner's official report. Few readers noted R. A. Statham's caustic reaction in the *Labor Advocate:* "I see that mine expert Jury of Coroner Brasher has made its report, and we want to say we were not disappointed in its findings. You have heard the saying Birds of a feather flock together. . . . They don't know when they may be in the same fix and may need an expert Jury."[9]

There were as well the statements that had been issued by national and state officials. As previously discussed, Joseph A. Holmes, head of the United States Bureau of Mines, arrived shortly after the explosion. He went into the Banner Mine and on emerging declared that it was one of the state's best-ventilated mines. Holmes declined to be specific: "I will say that from all indications it was a local explosion rather than the whole mine," he remarked. Holmes hastened to add that he did not get a chance to investigate closely. Ordered to the scene, James G. Oakley, president of the state Convict Board, had also investigated the Banner. Based on his brief excursion into the mine, Oakley made a prediction: the mine inspector's report would state that a local explosion had interfered with ventilation and brought on mass death through suffocation.[10]

Still missing were the official reports from the state mine inspector and from Washington. Yet a consensus had developed: the mine had been checked and declared safe before the day crew had entered. There were no gas concentrations or dusty conditions that could have precipitated an explosion. The Banner, according to the statement duly repeated for its implication that management was not at fault, was a model mine. To be doubly safe, the owners blamed human error in handling the bituminite for the explosion. Such an explanation separated the suspect shot firers, who were free, from the convicts. A standard objection to the utilization of convicts in mines was the use of an untrained and therefore dangerous labor force. If it seemed ironic that the experienced shot firers were to blame for the explosion rather than the convicts, there was the question of the greater interest to be served. It was the use of convict labor that constituted the volatile issue, and it was to the company's interest to resist any connection between the convicts and the explosion.

With the coroner's inquest, the death of Ernest Knight (and, through him, those of the other victims) was placed in the proper legal category. The social function had been performed, and death had been made more acceptable to the living by the ritual investigatory act. Even if Coroner Brasher had gotten the more intelligent decision that he hoped for, he fell far short of a technical investigation. That analysis was the proper function of the state mine inspectors, and there was some suspense and speculation as the days proceeded without a report.

"The official report will come from state officials," the *Montgomery*

Advertiser remarked, and a Birmingham paper agreed. No federal report would be binding. The editorials were correct. As the authority on the Bureau of Mines has written, "It was a research and education organization totally lacking in coercive capability."[11] In Washington, Holmes made no rush to release a report. Instead of mentioning the Banner specifically, the bureau chief commented on mine safety. In one speech he discussed various safety precautions to avoid mine explosions—wetting coal dust, proper storage of explosives, fireproof construction—but Alabamians wanted more details. At least some Alabamians did.

The real investigation by federal authorities had been accomplished by Dr. J. J. Rutledge, who remained behind in Birmingham. During part of his investigation, Rutledge was escorted through the Banner Mine by H. E. McCormack. By the middle of April, Rutledge was joined by George S. Rice, chief engineer of rescue men in the field. Rice came to Birmingham from his office in Pittsburgh. An official report was prepared, but no one expected it to conflict with the forthcoming statement from the state mine inspector. Holmes and his staff would decide whether to release their findings.[12] It was too early for the bureau to have established its credibility, and some miners and labor leaders were skeptical that it ever would. One critic well acquainted with coal operations wrote that Holmes was "a 'huge joke,' " and declared, "It is an insult to the intelligence of the miners of this country that men of the caliber of Holmes and Rice should be thrust upon them in running this Bureau of Mines."[13]

In his courageous rescue efforts at the Banner Mine, Chief Inspector Hillhouse had collapsed. Later he was sent to Johns Hopkins Hospital in Baltimore for an operation. The job of investigation fell on the assistant inspectors—Robert Neil and T. W. Dickinson. They examined the mine on April 15 and 18.[14]

As he waited for the reports, Governor O'Neal was also screening applicants to fill the positions provided for in the mine inspection law. Hillhouse, Neil, and Dickinson were engaged in their last official acts. On April 21, Inspector Neil made the trip to Montgomery to turn over his findings to the governor. Dickinson may have accompanied him, or more likely, he had already sent his analysis to O'Neal. Although the governor held Neil's report all day, he did not immediately release it to the press. Denied the document as a subject for news, reporters created news from the denial. Anticipation mounted to "tense interest," according to one journalist, who added, "mystery surrounds the findings of the mine inspector."[15]

Suspense and speculation were resolved the next day as the governor turned the Neil and Dickinson analyses over to the press. The treatment of the reports by the Birmingham newspapers proved instructive. The *Age-Herald* ran a long extract from Neil's report but devoted only two brief paragraphs to Dickinson's version of the disaster. The *News* managed to surpass even that degree of partiality by printing almost the entire Neil report. It alluded to Dickinson's views almost parenthetically and blurred the fact that the findings were diametrically opposed in their conclusions. Neil's statement was eminently acceptable to Pratt Consolidated; that of Dickinson was more troublesome because it raised major questions of blame against the company.[16]

Robert Neil produced a long and thorough analysis, obviously based on an extensive examination of the Banner Mine. It seems likely that Neil and Dickinson made their inspection trips together. Not until the twelfth left entry, 81 feet into the mine, said Neil, was there a sign of "a slight disturbance"—a fall of coal and rock. By the tenth left, at 342 feet, there was a minor disturbance, but the effects of heat were not discernible. At the ninth left entry 200 sticks of bituminite were found, along with 250 feet of fuse, 3 rolls of paper, and 6 sticks rolled up and ready for firing. At the eighth left entry an additional 75 sticks of bituminite were discovered. There was a great disturbance at this point, and even more by the seventh left entry. Here a box containing 4 sticks of bituminite, 25 feet of fuse, 4 rolls of paper, and 24 detonators was found. A stone and cement magazine had been built here, and it was here as well that "continuous falls of rock and coal" were encountered. There were, said Neil, indications of great heat in all the rooms of this entry. Rooms nine, ten, and eleven showed a "greater amount of explosive force than any other entry."[17]

Inspector Neil hoped that, by accounting for the original amount of bituminite in the Banner, he could ascertain the amount that might have exploded. According to Neil he found 206 sticks at nine left, 75 sticks at eight left, and 4 sticks at seven left—a total of 285 sticks, or 142.5 pounds of bituminite, that were recovered. Company officials, according to Neil, had stated that 160 sticks should have been in the magazine at seven left—40 left over from the night shift, 20 made ready for use, and 100 carried down by a shot firer on the fatal morning. Obviously the estimate of 160 sticks applied only to the magazine and not to the total bituminite in the mine. In fact, no one ever accounted for all the bituminite that Neil discovered.

Up to this point in his report, Neil was clear and easy to follow. But at his next step of reasoning the matter clouded. "Further evidence," wrote Neil, "shows that 120 sticks of bituminite were missing at the magazine." That was a curious conclusion, since Neil's only mention of bituminite in seven left were the 4 sticks found in the box. Neil did not suggest that shot firers could have taken some of the 160 sticks into other entries. Nor did he explain why he thought 120 sticks of bituminite were missing rather than the obvious total of 156 sticks. At any rate, Neil was postulating a bituminite explosion of at least 120 sticks, or sixty pounds.[18]

It seems now that the argument from mathematics was less than conclusive. Yet it formed the basic evidence for Neil's conclusions. Missing from the magazine were 120 sticks of bituminite. The shot firer who carried 100 sticks was found near the magazine badly mutilated. The automatic fan recorder indicated no interruption in the ventilation system, and, Neil argued, there had never been much gas in the Banner. The night fire boss had found all rooms clear of gas except for a small amount in nine left that was harmless. Therefore, Neil arrived at a clear conclusion: "The explosion originated by a premature explosion of bituminite at the magazine." There might be speculation on why the bituminite exploded, "but the fact remains that it did explode, as it was not found at or near the magazine." Neil admitted that the explosion might have been "aided by dust," but he stood firm on the point of a premature explosion.[19] Neil's investigation contained nothing to worry Pratt Consolidated. The explosion was not the product of gas; it was not originated by a coal dust explosion; it was not the fault of the ventilation system. Human error, probably by a shot firer, had caused the disaster.

The report of Inspector T. W. Dickinson told a completely different story. He agreed with Neil that the force of the explosion had divided at room seventeen near the magazine, but he believed that it had originated at either room thirteen or fourteen. Dickinson cited the company's explanation that the bituminite had exploded prematurely—and that, of course, was Neil's conclusion as well—but he placed no credence in that judgment. "I examined the place very carefully," wrote Dickinson, "but failed to find any indication of any powder having been exploded, either there or elsewhere in the mine." The disaster "was caused by gas having been ignited in either room 13 or 14 on the seventh left heading, the same generating enough heat to ignite the dry dust with which it came in contact, thus propagating and feeding the force of the explosion."[20]

Had Neil and Dickinson examined the same mine? Were they reporting on the same disaster? It was a badly split decision. A Montgomery newspaper wondered who was right: was it Neil (the explosion originated from a premature explosion of bituminite, probably aided by dust) or was it Dickinson (the explosion was caused by gas)? Other journals were also curious about the conflicting reports. The *Birmingham Ledger* correctly noted that Neil's investigation held no one responsible. The *Ledger* mentioned further that Hillhouse, recovering from an operation, had written his part of the report.[21] Whether he aided either or both inspectors was not made clear. There is no real evidence that the ailing chief inspector contributed at all.

Nothing was said in Neil's statement and presumably not in Dickinson's regarding the question of fan ventilation in the mine. Only the larger fan was actually in use at the time of the explosion, since the second and smaller fan was used for emergency, back-up service. It seems plain that neither fan was fitted with explosion doors designed to relieve the pressure and thus save the fan and fan house. It must be assumed that the explosion knocked out not only the main fan that was running but also the small fan. One miner stated the obvious when he wrote, "If the fan could resume work immediately after an explosion, rescue parties could at once proceed to explore the mine and give relief to the injured."[22]

As it turned out, Dr. Holmes filed a thorough report. It was compiled by his staff and based entirely on their investigations. Dr. Rutledge and Rice collected nine samples of gas from the mine and forwarded them to Washington on April 25. Following laboratory analysis, the presence of explosive gas was established. Other evidence was assembled following a letter by Dr. Holmes to Dr. Walter S. Roundtree of Wylam, Alabama, who had assisted in the rescue attempts. The head of the Bureau of Mines wanted a description of Dr. Roundtree's symptoms while in the mine and what the aftereffects were. The Alabama physician replied on May 10 (and included testimony from three other participants). Dizziness, rapid loss of strength, and labored breathing were symptoms common to all. The men experienced headaches, vomiting, and general debilitation that lingered for ten days to three weeks.

The preliminary findings were not completed until September. In the published report, the investigators drew substantially the same conclusion as that of Assistant Mine Inspector Dickinson. According to Dr. Rutledge, "Gas had been found in several places by firebosses that morning [April 8]; it was concluded that an unreported accumulation in an idle room was ignited by the open lamp of a miner. The explosion was propagated by gas and coal dust."[23] Although a miner inadvertently triggered the explosion, the actual cause, as Rutledge pointed out, was an unreported accumulation of gas. Had the gas not been present, there would have been no explosion.

Such objective information as that supplied by the Bureau of Mines would have been damaging to Pratt Consolidated in a court of law.

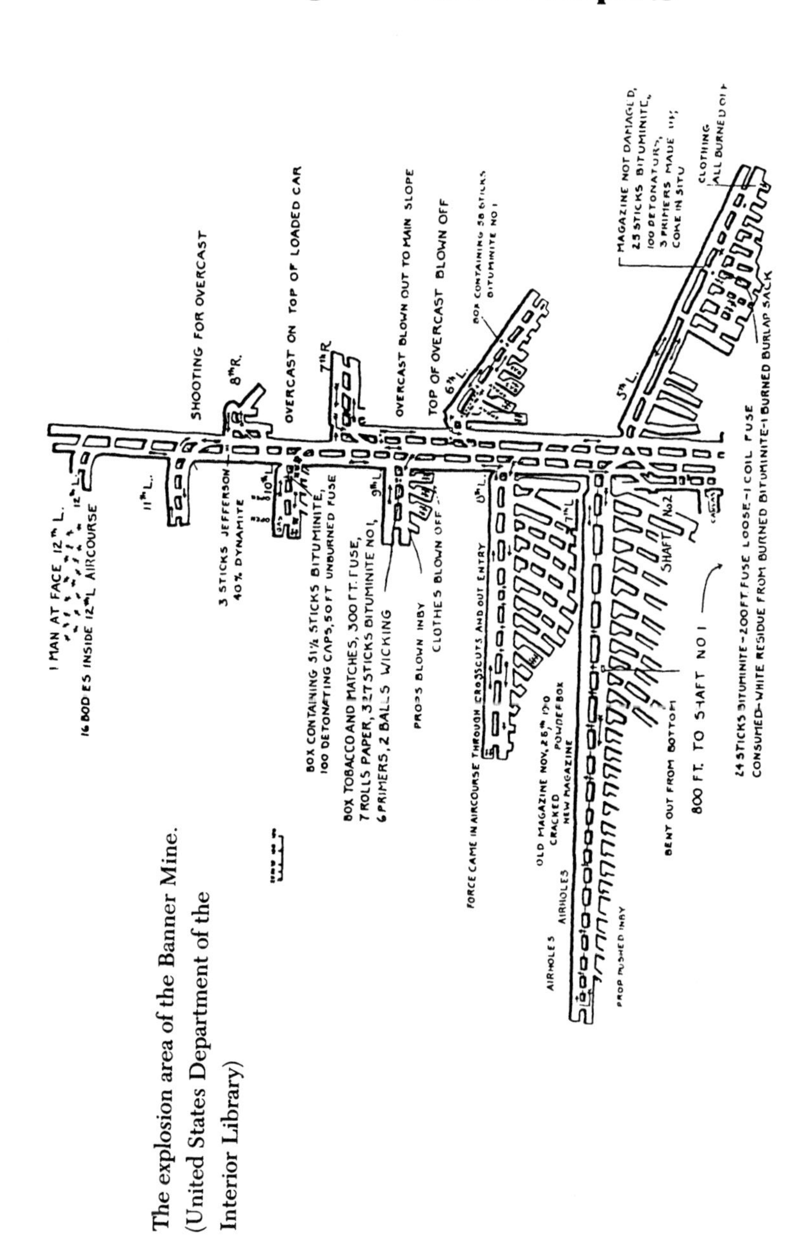

The explosion area of the Banner Mine. (United States Department of the Interior Library)

Yet no newspaper mentioned the federal report, and it was not filed in the governor's papers. Assembling the data and putting it in published form could not have occurred until late 1911 or, more likely, 1912. By then no one took notice. The accident was not a pressing issue.

Governor O'Neal, having received the state mine inspectors' reports but not that of the Bureau of Mines, next moved to fill the seven slots created by the new mining law (a chief inspector and six associate mine inspectors) as well as that of mining expert (or mining engineer). There was some pressure to reappoint Hillhouse as chief inspector. The ailing Hillhouse was praised for his good work "in all the fearful explosions of this district," but the governor was determined to name an entirely new set of officials.[24]

Before the names were released (April 22–23), Jesse F. Stallings—weathered by the wild exigencies of Alabama politics and a shrewd judge of men—offered the governor some gratuitous advice. Stallings was well acquainted with mining at the technical level and with miners and labor leaders at the personal level. Knowing the proposed new officials, Stallings declared them "incompetent," men whose appointment "will create, in my judgment, a storm of indignation throughout the district among the twenty-two thousand toilers in the mines. I earnestly ask you to hear the miners—the other side and appoint competent men."[25] Speaking for UMW District 20, President J. R. Kennamer asked for men partisan to labor and specifically requested the appointment of T. A. Tinney and Frank Hillman.[26]

Without doubt O'Neal believed that he was appointing competent men. That his selections were, as Stallings predicted, more acceptable to the operators than to the miners came as no surprise. Management was well pleased when Charles H. Nesbitt of Jefferson County was named chief inspector. The thirty-eight-year-old Nesbitt, a native Alabamian born in Cherokee County, was described as "one of the best known young mining engineers of North Alabama." Nesbitt's straight brown hair and thin face accentuated by a long jaw combined to give him a look of great earnestness. At the time of his appointment Nesbitt was resident engineer of Sloss-Sheffield Steel and Iron Company mines at Maben in Jefferson County. The other officials, all from mining district counties, were Mining Expert C. C. Huckabee and Associate Inspectors Judson Franklin Webb, W. W. Kicker, W. R. Ray, Thomas Roscoe, David Kelso, and Frank Hillman. Huckabee was a former vice president of the Yolande Coal and Coke Company.[27]

The state press, with few exceptions, joined the operators in approving of O'Neal's appointments. While the miners were less than satisfied, the "storm of indignation" predicted by Stallings failed to materialize. In Hillman, a resident of America in Walker County, the miners obtained a sympathetic representative. Nesbitt and the associate inspectors possessed adequate credentials. Their job performance remained to be tested, and to his credit, Nesbitt bestirred himself and his staff. They launched a round of visits to all of the state's coal mines.[28] In any case, the miners were too disorganized to make an effective demonstration of their feelings.

Once the reports of Neil and Dickinson were submitted, state officials laid a protective blanket over the episode. In the company of Assistant Superintendent W. Asa Gibson, Nesbitt visited the Banner in May. Prior even to making his investigation, Nesbitt said, "The reports of the state mine inspectors as to conditions before the disaster showed the place to be in good condition, so far as was discernible." Apparently, Nesbitt had not read Dickinson's report carefully, and he may not have been furnished with Rutledge's findings. Nor had prison head Oakley done any significant homework. He examined the Banner again in May and announced that there was no reason for censure. Pratt Consolidated had done everything in its power to care for the convicts. The explosion was unpredictable, and in a selective and subjective analysis of the reports, Oakley said, "According to the best of the experts on the subject, [the Banner] had been kept in the best of shape possible." Len F. Greer, associate convict inspector, echoed the earlier general statements of President Maben of Sloss-Sheffield. Greer looked on life at the Banner Mine as being pleasant. The prisoners were provided with a daily shower and a change of clothing. Their beds were comfortable and their quarters well ventilated. Beyond that, the men earned money to send home by performing extra tasks.[29]

The appraisals by Greer and his superiors were remarkable for what they failed to reveal. As 1911 ended and 1912 began, the most unlettered prisoner, whether convicted by the state or a county, understood what his duties would be when he was sent to the Banner Mine. He knew also that he was to behave as ordered but was unacquainted with the carefully structured set of rules and regulations that governed him. The world of any prisoner was tightly hedged by words as well as by locks and guards. Alabama's three-man Board of Inspectors of Convicts—Oakley; J. T. Fowler, physician inspector; and L. F. Greer, associate inspector—enlisted the services of the state printers and binders and published a revised set of ordinances.

Part 1 outlined the duties of the president, Board of Inspectors, chaplain, physician, and clerks. Rules for the contractors and wardens were established by part 2, while part 3 contained special regulations for mines. Other divisions dealt with county convicts, officers and employers, discipline, punishment, purchasing agent and auditor, hospital and sanitation, tuberculosis hospital, descriptive record, and miscellaneous. Altogether there were 145 rules and regulations.

Regarding mines (rules 43–45) there were strict requirements concerning ventilation, drainage, and entrances. There would be a mine foreman and his assistants and an engineer. The deportment required of convicts was spelled out ("obey strictly all orders of the Mine Foreman, or his assistant"). All prisoners were to have a "good set of mining tools."

The general sections applying to discipline and punishment were harsh. Punishment was not to exceed fifteen lashes, "to be administered over the clothing usually worn by the person." Extreme cases—attempts to escape, "persistent insubordination, or outrageous conduct"—merited twenty-one lashes. Solitary confinement on a diet of bread and water was permitted. Such confinement was not to exceed thirty days, and after the third day, a daily examination by a physician was required. The doctor was to determine whether the punishment should be modified or terminated. Such were the rules that applied to the prison force of Alabama.[30]

Official absolution aside, Pratt Consolidated was soon confronted with a series of court cases. Even though the state declined to intervene on behalf of those killed at the Banner Mine, it was inevitable that private suits would be entered. Within two weeks after the explosion, the initial damage suit was filed in city court. As "his next friend," Sallie Smith demanded ten thousand dollars in damages for Jesse Smith (race not given). The claim was based on severe injuries, including deafness, that Smith suffered when the accident occurred. On May 10, the first cases resulting from actual deaths appeared on the city court's docket. H. W. Wright sued as administrator for the estates of black convicts Dolphus Rogers of Morgan County and Will Garth of Jefferson County. Lawyers for Wright claimed that Pratt Consolidated had permitted gas or dust or both to accumulate in the Banner. The explosion that followed was blamed on the company's negligence, and ten thousand dollars in damages was asked in each instance.[31]

During the ensuing months, numerous cases were heard in both the city court and the circuit court. Usually, the plaintiffs asked for damages ranging between five thousand and twenty thousand dollars. Birmingham's newspapers—journals of record as well as of news—attempted to cover at least some of the trials, but the suits attracted no attention on their editorial pages. Yet the magnitude of the accident meant that a series of adverse decisions would injure Pratt Consolidated's public image and would prove financially costly. The corporation moved to protect itself by bringing onto the scene the city's finest lawyers, particularly R. B. Watts.

Some city and circuit court cases were concluded at their point of origin, but the vast majority were appealed to a federal tribunal, the Circuit Court of the United States of America, Northern District of Alabama, Southern Division. The law involving civil suits for damages declared that, when a plaintiff sought damages of two thousand dollars or more, the defendant, if a citizen of another state, could refuse to be sued in a state court. Pratt Consolidated was incorporated in Delaware and, as Watts argued, was not a citizen of Alabama. The corporation most certainly did not wish to rest its fare in the hands of a local jury. Since all of the plaintiffs demanded jury trials, the corporation figured correctly that sympathy would lie with the dead miners. Demurrers were filed, and the cases were appealed to federal court.

Even so, there was some action at the lower bars of justice. A case involving Daddy Denson (a free black worker) never reached Circuit Judge John C. Pugh. Instead, a settlement was made out of court for an undisclosed amount. Suits for two other black convicts who died at the Banner ended in a similar dismissal by the circuit court—"satisfactorily adjusted by and between the parties."[32] Seven cases ended in city court, with no definitive accounting. Still, in three cases heard in city court, settlements were made that awarded one hundred dollars each to plaintiffs for the injured Jesse Smith and the deceased John Lovelace and Will Williams.[33]

Twice in city court and once in circuit court, plaintiffs sued for $1,999. The amount sought denied Pratt Consolidated the opportunity of appealing to federal court. Mattie Bowers, as plaintiff for Levie (Levy) Bowers, a black convict killed at the Banner, sued for $1,999 in city court. Julie Brown, administrator for the estate of the deceased black convict, Will Brown, sought a similar judgment. The record contains no final disposition of either case. In the circuit court Harriet Phillips, plaintiff for Eddie Grimet, a black prisoner killed at the Banner, sued for $1,999. The settlement for Phillips is not a matter of record.[34]

The wisdom of Pratt Consolidated's adroit legal ploy was revealed when at least fifty-one suits were appealed to federal court. In only six cases was a decision rendered for the plaintiff. The litigant for the estate of John Wright was awarded three hundred dollars, the largest amount. Boyd Love's administrator received two hundred dollars, and one hundred dollars each were awarded to four other plaintiffs. Pratt Consolidated had to pay a total of only nine hundred dollars in federal court (and was usually assessed court costs as well).

The higher tribunal dismissed a few cases because there were no lawyers present to prosecute them, and one suit was thrown out because it was erroneously entered. The majority were dismissed because a satisfactory arrangement was made between the company and the plaintiffs. Because the cases were not heard, the amount of the settlement was not revealed.[35] On the basis of amounts awarded in cases settled without trial by city and circuit courts and reported in newspapers, we can assume that settlements ranged from one hundred to three hundred dollars. It is probable that in some cases the company paid nothing at all.

For all its fears, Pratt Consolidated had an easy time in the courts. Whether the result of private arrangements or of decisions rendered by city, circuit, and federal courts, financial punishment for the corporation was minuscule.

Officials at Pratt Consolidated greeted the late summer of 1911 with relief. The Board of Revenue did not cancel its contract. Without missing any revenue, the board collected quarterly earnings from the company in June amounting to $4,319 (there were additional payments for individual prisoners).[36] It was good to have the noncondemnatory report from Assistant Mine Inspector Neil, while that of his colleague Dickinson, with all its potential for trouble, had caused little negative reaction. State officials were well disposed toward the corporation, and the investigation by the Bureau of Mines never received general circulation. In the courtroom, plaintiffs had fared badly in their attempts to collect damages from the company. A strike by the impotent UMW was inconceivable, and the mine safety law had hushed many critics. Prisoners assured a steady supply of labor and, as required by convict department regulations, an obedient one.

In July the Banner had settled back into its preexplosion routine. To demonstrate the efficiency of their operation, officials of Pratt Consolidated staged a Saturday barbecue and invited state, county (including the Board of Revenue), and city officials. The visitors inspected the mine and observed its operation. For a brief interval the attraction of pork meat cooked long and basted with pungent sauces made equals of officials, politicians, and convicts. The occasion was a sham, a labor journal declared. A picnic could not erase the past. Reminding its readers that the Banner tragedy was history by less than three months, the paper denounced the company's method of handling convicts with "barbecue and explosives."[37]

The demand for coal was high, and the company moved to increase production. All the Banner needed was more convicts, but as the *Birmingham News* reported, "Not until the courts are grinding will there be as much labor in the Banner mine as is needed."[38] No better arrangement seemed possible, unless the state offered to produce coal (with convict labor) and to sell it to Pratt Consolidated. To the surprise of McCormack, Ramsay, and others, that was exactly what happened. As president of the Convict Board, James Oakley had devised such a proposal and had persuaded Governor O'Neal that it was valid.

7
"The Moral Question Obtrudes"

Shortly after he was appointed president of the state Convict Board in March 1911, Oakley devised a bold and controversial plan. He convinced O'Neal that state revenues would increase if convicts from south Alabama's lumber camps were concentrated in the mines of the Birmingham district. Oakley's main argument was that expenses would be dramatically reduced. After consulting with various advisers and experts, O'Neal agreed. Oakley then advised the timber contractors that their leases would be canceled in November 1911. When the lumber interests protested, it was agreed to permit some of the contracts to continue for a limited time. In that way the state would gain revenue from those convicts physically unable to work in the mines. As Oakley told the governor, the lumber companies "would use without any classification any kind of convicts he might furnish, such as were known as peg-legs, cripples, etc."[1] The able-bodied prisoners would go to the mines.

Oakley entered negotiations with McCormack. When the question was raised whether the state was receiving sufficient compensation, O'Neal held conferences with state Mine Inspector Nesbitt, state Mining Engineer Huckabee, and others. O'Neal finally gave his approval. Oakley and McCormack signed the contract on August 5; it was to extend from November 15, 1911, to November 15, 1916. It would not go into effect until January 1, so that O'Neal could investigate all of the ramifications.

Among those consulted by O'Neal was Jesse Stallings. As usual, Stallings was candid, and he was also skeptical: Oakley was "inexperienced in mining affairs, and [I] feel that he has been overreached in his contract leasing the Banner mine." Stallings carefully investigated the contract's terms and bluntly wrote O'Neal that the result would be "a loss to the State and an injury and injustice to your administration."[2] The governor also appointed a special commission to look into the matter. Following much consultation and after adjustments and concessions by all parties, the contract went into force.

The agreement was an extremely detailed document that ran to nineteen clauses, and it represented a major change. The state agreed to use four hundred prisoners and operate the Banner Mine. Pratt Consolidated would purchase the entire output, which would be delivered on railroad cars outside the stockade. The workers would be state convicts, although the company was permitted to transfer its county prisoners to other property. By doubling the number of prisoners at the Banner, production would increase from 1,100 tons of coal a day to 1,800, perhaps to 2,000. The state stood to make sixteen thousand dollars a month. Both parties accepted numerous specific obligations, but the state had paramount power in that it could terminate the contract at any time without giving a reason.[3]

Pratt officials (including W. Asa Gibson, manager of convicts for the company, who helped negotiate the contract) were delighted. Oakley emphasized that the state was not going into open market coal mining and invited other companies to make similar arrangements. Yet some owners interpreted the deal as undisguised favoritism. L. B. Musgrove, president of Corona Coal and Iron Company, spoke his mind: "The state policy should be against entering the commercial coal field and competing with commercial operators." Musgrove voiced the sentiments of other operators. "When the state leases a commercial mine and operates it with convict labor, those of us who are engaged in that branch of mining feel this cheap labor competition very seriously. The state's policy in this matter is emphatically wrong."[4]

Although TCI withdrew from leasing convicts, it did so grudgingly and without any outcry of moral indignation against the system. In

1907 the corporation had been taken over in a celebrated transaction by United States Steel. George Gordon Crawford, a native Georgian and an experienced steel executive, was installed as president of TCI. Crawford became instrumental in launching the corporation on an ambitious program of "welfare capitalism." Without prison labor as insurance against strikes and strongly opposed to union influence among its workers, the company adopted a policy of cooperation between management and labor. Social, recreational, and educational programs followed, and there was special emphasis placed on upgrading the workers' health.[5]

If TCI moved away from leasing, Sloss-Sheffield Steel and Iron remained active. That corporation's contract was due to expire in December 1911. Oakley informed O'Neal of the pending termination in August, and negotiations were begun. Company officials complained of previous discrimination. The state agreed, and a settlement was reached: Alabama's convicts would be leased at a price equal to that being paid for free labor. There were obvious flaws in such an arrangement. Even if, at a given time, equal pay arrangements were worked out, a constant would rarely match a variable—or worse, the variable (the price for free labor) would be artificially kept down to conform to the agreed-upon pay scale. After a number of compromises, the contract was signed on March 30, 1912.[6]

Although the contract between the state and Pratt Consolidated had been signed in August and became operative in November, it was not until January 1, 1912, that the transfer of prisoners began. Governor O'Neal had delayed until he was certain that the arrangement would work. A few men arrived at the Banner separately, but the majority were transferred by train from the property of TCI. The move was carried out under the direction of Mining Expert Huckabee.

The exchange was made without incident, although as a spectacle, it attracted curious onlookers along the way. On New Year's Day, three hundred convicts, handcuffed and chained two by two, were marched in groups of fifty from the stockade at TCI mine number twelve to a waiting train. The prisoners were placed in five cars, and their manacles were removed. Each car was guarded by four men—two inside, one at each platform. They were armed with pistols, shotguns, and rifles. Further security was added by the presence of sixteen bloodhounds, including the five-year-old bitch "Old Dutch," renowned for her tracking abilities. A reporter noted that the convicts, including twenty-five whites, seemed to be in good spirits. It was characteristic of their race, according to the reporter, that the blacks carried various musical instruments, including banjos. "Another characteristic which a few of them showed," he wrote, "was a superstitious fear of going into a mine where 128 convicts had been killed by accident last year."[7]

From all outward appearances, the Banner tragedy had diminished neither the profits nor the prestige of Pratt Consolidated. An exclusive contract with the state for producing coal at the Banner Mine and the right to continue leasing county convicts at its other properties seemed to assure future profits. The state's own statistics would bear out the efficacy of convict leasing. From 1910 to 1914, Alabama's prison system yielded a profit of $2,188,604.88. The Banner more than provided its share of revenue. Between January 1, 1912, and September 1, 1914, state profits from the Banner Mine amounted to $300,149.[8] Despite the impressive figures, all was not well with the system. As one Birmingham paper, not known for its opposition to management, remarked, "the moral question obtrudes."[9]

The state's safety, health, and security record after taking over coal production at the Banner Mine was less than impressive. There were no major explosions, but on December 7, 1912, a fire destroyed the tipple, washer, and coal bin. Fortunately for Pratt Consolidated and more fortunately for the prisoners, the fire occurred at three o'clock in the morning when the mine was empty. The damages were estimated to be forty thousand dollars. Between 1912 and 1914, six convicts at the Banner were crushed to death by coal cars, two were killed by falling rock, one was "killed in mine," ten died from diseases, and one burned to death. In addition, six were murdered by other prisoners, one was poisoned, and another fatality had no

Aftermath of a fire at the Banner Mine, December 7, 1912. (Birmingham Public Library Archives)

known cause. All of the victims were blacks. The only white convict to die during the period drowned accidentally.[10]

The new death list and the grisly record of the past stirred various groups in Birmingham to take action. As agents of conscience and image, they were partially successful in altering their area's reputation as a "dumping ground" for prisoners. Birmingham newspapers joined the fight. After TCI ceased leasing, the corporation also became an opponent of the system. By 1913, the withdrawal of TCI sharply reduced county revenue. Pressure mounted for more and better roads as the number of automobiles increased yearly. If highways were inevitable, they were also expensive. In April 1913, the Board of Revenue began assigning all new able-bodied convicts to road work. The experiment paid off: money saved in road construction more than offset money lost from leasing. The necessary amount of public indignation, coupled with an economic alternative, thus brought an end to convict leasing in Jefferson County.

The change in policy produced a number of skeptics, and George B. McCormack was among the unconvinced. He continued to insist that it was preferable to employ convicts in coal mines, that it was "infinitely better from every standpoint of health, humanity, and economy than working on the roads."[11] Other counties continued to send prisoners to the Birmingham area, and the state's leasing contracts remained in force. But Jefferson County convicts never returned to the mines.

Scandals at the state level concerning leasing contracts, especially at the Banner Mine, and dishonesty within the convict department hurt the O'Neal administration and focused attention on the leasing system. Oakley's two-year term ended in 1913, but before reappointing him, O'Neal decided that a closer look at the convict department was warranted. Revenue from the Banner Mine and other commitments (the textile mill at Speigner), while substantial, were not up to expectations. At the governor's orders, state examiners Frank V. Evans and H. Y. Brooke checked the records of Oakley and Theo Lacy, his chief clerk. The examiners uncovered a scandal of major proportions.

A law passed in 1907 had given the president of the Board of Convicts control over the department's expenditures and receipts with a minimum of accountability. With their belated inquiry, Evans and Brooke discovered discrepancies amounting to an estimated $150,000. There were strong indications that the scales used to weigh coal at the Banner had been tampered with to the state's loss.[12]

As a result of the public investigation launched in March 1913, Oakley was removed from office, arrested, and charged with embezzlement of state funds. There were accusations of misfeasance and malfeasance in executing contracts for the hiring of prisoners. When Lacy disappeared with ninety thousand dollars, the defalcation added to the scandal. Lacy voluntarily surrendered ten months later and, in March 1914, was convicted of embezzlement and sent to prison for ten years. A few months later another trial and conviction added six more years to Lacy's sentence. Oakley was also tried for embezzlement in March but, to the dismay of many, was found not guilty. Having been found innocent of the state's charges, Oakley was tried at a more local level. Again he was acquitted, this time by a Shelby County jury, and he never served a day behind bars. He thus escaped the irony (and agony) of becoming a victim to the convict lease system that he had championed and administered, however badly. Governor O'Neal did not escape unscathed, although charges and innuendoes that he was implicated were false and largely politically motivated. At the least he appeared guilty of poor administration, of allowing a situation tailor-made for corruption to get out of hand.[13]

Despite public exposés and criticism, the state remained in the business of hiring out its prisoners. The basic reason remained the same. As the embattled Robert Moulthrop said, "There is but one argument in favor of working the convict in the death-dealing coal mine, and the malaria-breeding district where the new saw mill is established, and that is the revenue derived from the lease system."[14]

Governor Charles Henderson (1915–1919) presided over the same sordid arrangement and made no attempts at reform. Thomas E. Kilby (1920–1924) succeeded Henderson and became the first chief executive to push for meaningful change. In 1919, the legislature responded to Kilby's leadership. It passed one measure that modernized the penal structure, and it replaced the state Board of Convict Inspectors with the state Board of Control and Economy. Another law ended leasing after January 21, 1923. As it turned out, the latter act did not specify that convicts could not be worked in the mines, only that they could not be leased. To make matters worse, a special session of the legislature in 1921 moved the date forward to January 1, 1924.[15]

Critics redoubled their efforts. A series of damning articles appeared in national magazines, more scandals developed in Alabama prisons, and events in Florida (unsavory as they were) had a positive effect. Martin Tabert, a young North Dakotan, was arrested in Florida and later beaten to death in one of that state's notorious prison camps. The case made headlines across the nation, and the public outcry that followed forced Florida to abandon convict leasing in 1923.[16]

The publicity from the Tabert episode crossed the state line, aroused the consciousness of Alabamians, and pushed them to copy Florida's example. Under Governor William W. ("Plain Bill") Brandon (1923–1927), an act set March 1927 as the date on which to end the system. Then the effective date was extended still another year, and the system continued until the administration of Governor Bibb Graves (1927–1931). Even then there were interests that wished to preserve it, but the governor yielded to extreme pressure. He signed a law setting the terminal date of June 30, 1928. After that time no prisoners could be leased. Specifically, it became "unlawful to work any convict, State or County, in any coal mine of Alabama."[17] Seventeen years had passed since 1911. George B. McCormack had been dead since 1925, the same year that Pratt Consolidated merged with Alabama By-Products Corporation, which took controlling interest.[18]

Almost two decades had elapsed since L & N agent J. Flavious Erwin, walking routinely along in a soft April drizzle, heard the muted sounds of an underground explosion in the Banner Mine. The convicts who perished had few to mourn them and few to champion a cause that would never have allowed them to be in a coal mine. Everybody agreed that it was unfortunate for 128 men to die. But industrial accidents were bound to occur, despite all efforts to prevent them. The dead convicts were soon forgotten as others took their places. As Hartwell Douglas, chief clerk of the Board of Inspectors of Convicts, explained to Governor O'Neal, "The argument that in coal mining [prisoners] come in competition with free labor seems far-fetched, for the reason that in . . . any other labor the same competition exists." For Douglas it was all elementary: "Convict labor is desirable by firms or corporations for the reason that it can be depended on—always ready for work—contracts for the output can be made with the knowledge that the goods can be delivered."[19]

Why had reform failed in 1911 when the age-old iniquities of the lease system were highlighted by the drama of mass death? There were the usual reasons of time and circumstance. In 1911 the issue had been in abeyance, a lull between the last great effort and the next period of moral outrage. As the culmination of a campaign for abolishment of the lease system, the Banner disaster might have provided the last necessary emotional element for success. As the first incident, however, so easy to dismiss as accident and not inherent to an evil system, it burned itself out too rapidly to ignite a long-term assault on the penal establishment. In addition, the explosion at the Banner raised not one issue but two. The demand for higher standards of mine safety challenged the interests of the mine owners. Yet that public issue could be answered to some degree by the politicians without doing damage to their own vital interests. The mine safety issue split the prevailing establishment just enough to allow one step of progress, a shortened step, to save the coal companies from a full measure of responsibility. On the convict lease question there was no split, and there was no change.

The position of Governor O'Neal was a perfect illustration of reform meted out by the pressure of necessity. To his credit, O'Neal fought hard to obtain passage of the measure reforming conditions in the coal mines. Yet for him the mine safety law was a sufficient response. He thus declined to lead the fight to abolish the lease system, a position calling for an attack on powerful economic interests and for creative statesmanship. It remained for Senator Moulthrop, bitter in his defeat, to read the future.

If the Alabama convict lease system had functioned simply as a penal establishment, as a method of incarcerating and utilizing the bona fide miscreants and the vicious of society, it would still have been open to serious questions. Inevitably, it would have been the target for penal reform. But the system, in the name of dealing with criminals, created and fostered the use of forced labor. It functioned fully as much—and sometimes more—to supply the requisite cheap labor to favored interests as it did to handle criminals. In a sense the Alabama Black Codes of Reconstruction were never repealed. The social control and the economic reward they produced were now narrowed in scope but unchanged in method. A black arrested in 1866 and charged with vagrancy was usually compelled to work out his sentence in forced labor on a farm. Yet he was no different from the black arrested in 1907 on a gaming charge and sentenced to hard labor in a coal mine so that the county could meet its obligations to a mining company. Convict leasing was the Black Code of the New South, no longer necessary for mass control (although the potential was always there), still masquerading as a part of the legal system and law enforcement. It served government and industry to the great reward of both.

That was why the convict lease system was an evil one, and that was why it was finally abolished.

Appendix:
Prisoners and Free Workers Killed at the Banner Mine Explosion

Contemporary accounts of the accident, both official and unofficial, listed the dead at 128. The authors have used that figure in the text, but, as indicated by the list below, believe the number was 129. It is also probable that there were several unreported deaths occurring later that could be attributed to the explosion.

Prisoners by County

Barbour County
Son Dorson
Minn McDaniel

Butler County
George Broughton
George Gilmore
John Howard (white)
Will Sims
Garfield Smith

Calhoun County
Ed Causey (white)
Ernest Leach

Chilton County
Sydney Dunn (white)

Escambia County
Claude Henry
John Lovelace
Bee Linquirt
Mack Pounds
Robert Seals
Will Williams

Franklin County
Antney Hamilton

Greene County
Logie Croxton
Robert Gardner
Essex (or Sip) Hill
Blake Hinton

Appendix

Jim Jackson
George Mobley
Douglas Powell
Henry Wyser

Henry County
Son Lewis

Jefferson County
George Adams
Gus Ammonds
Dillard Battle
Levy Bowers
Lee Bowie
Will Brooks
Joe Brown
John Brown
Joe Butler
John Campbell
Tom Chowchow (white)
Jeff Cobb
Bester Collins
Dusty Cook
Tom Cosby
Sam Daniels
Comer Davis
Dan Davis
James Davis
Will Davis
Sam Echols
Will Evans
William Garth
Charles Graham
Henry Grimmett
Charles Harris No. 1
Charles Harris No. 2
Bud Hicks
Abe Holly
Ed Jackson
Jim Johnson
Johnnie Johnson
Will Johnson
Henry Jones
Jim Kidd
Ernest Knight
Kill Kutton
Ed Kyatt
George Lawson
Boyd Love
Judge Luckey
Haywood McCoy
Sam Mathews
Lonnie Meers
John Moore
Lyden Moore
Sonnie Morrow
Robert Nelson
Will Nepples (Nipples)
Josh Parish
Alex Pope
Jim Redd (alias Hurd)
Sim Robinson
Frank Sanders
Budd Savage
Will Shaw
Oliver Singleton
Wesley Smith
Herman Spencer
Elijah Strickland
Fate Tiller
Charles Turner
John H. Walker
Ed Warner
Arthur Welck
James Wheeler

Dee Willer
Ben Williams
George Williams
John Wright (white)
John Wyley

Marion County
Frank Ladro (white)
Charles McDonald (white)
Ike Taylor

Morgan County
Pollard Abernathy
George Featherstone
Arch Hill
Prentice Johnson (white)
Frank Pruitt
Dolphus Rogers

Perry County
Top Brooks
Joe Lowery
Columbus Nave
George Sanders

Tallapoosa County
Berto Pearson
George Washington

Tuscaloosa County
Robert Bailey
George Green
John Hill
John Hy Johnson
Henry Lipscomb
Willie Moore
Richard Myree
Gus Thornton

Walker County
W. G. Browner (white)
Sam Mason
Bat Waters
F. Williams (white)

Free Workers
Daddy Denson
Lee Jones (white)
Mose Lockett
O. W. Spradling (white)
Dave Wing

Notes

Introduction

1. Carl Carmer, *Stars Fell on Alabama* (New York: Farrar & Rinehart, 1934), p. 81.

2. David Brody, *Workers in Industrial America: Essays on the Twentieth Century Struggle* (New York: Oxford University Press, 1980), p. 4.

Chapter 1. Death at the Banner

1. Descriptions of the explosion and its aftermath lean heavily on the *Birmingham News*, April 8–17, 1911, and the *Birmingham Age-Herald*, April 9–10, 1911. Other local newspapers, including the *Birmingham Labor Advocate*, April 14, 1911, were useful. On various points wire service reports were good (although they depended on the value judgments of rewrite men and on space requirements). See the April 9, 1911, issues of *New York Times, Atlanta Constitution, Macon* [Georgia] *Daily Telegraph*, and *Augusta* [Georgia] *Chronicle*.

2. For a description of the Banner prison and mine, see *Birmingham Howle's Iconoclast*, April 22, 1911, and the details of prison life given in *Quadrennial Report of the Board of Inspectors of Convicts for the State of Alabama to the Governor, September 1, 1910–August 31st, 1914* (Montgomery: Brown Printing, 1915).

3. See Robert Neil to Governor Emmet O'Neal, *Birmingham News*, April 23, 1911.

4. *Northport West Alabama Breeze*, April 19, 1911.

5. *Birmingham News*, April 8, 1911.

6. *Birmingham Ledger*, April 10, 1911.

7. *Birmingham News*, April 8, 1911. The *Tuscaloosa News*, April 11, 1911, reported that, because the regular telegraph line between the Banner and offices of the Pratt Consolidated was down, Erwin used the L & N line as the only means of communication.

8. *Savannah* [Georgia] *Morning News*, April 9, 1911. For a description and pictures of the mine safety cars, see Herbert M. Wilson and Albert H. Fay, *First National Mine Safety Demonstration* (Washington, D.C.: Government Printing Office, 1912), pp. 27–28. Birmingham was listed as a mine safety station in 1911. By 1912 the mine safety cars had been reshuffled and car no. 7 served Kentucky, West Virginia, Tennessee, and Alabama.

9. For Hillhouse see *Birmingham News*, April 17, 1911, and *Birmingham Age-Herald*, April 23, 1911. For Dickinson see *Birmingham News*, May 17, 1910.

10. Ethel M. Armes, *The Story of Coal and Iron in Alabama* (Birmingham: Chamber of Commerce, 1910), p. 393. Thomas M. Owen, *History of Alabama and Dictionary of Alabama Biography*, vol. 4 (Chicago: Clarke Publishing, 1921), p. 1099; George M. Cruikshank, *A History of Birmingham and Its Environs* (New York: Lewis Publishing, 1920), p. 80. See also Erskine Ramsay, "Autobiographical Statement," pp. 75–79, Erskine Ramsay papers, Archives Division, Birmingham Public Library, Birmingham, Alabama.

11. James Saxon Childers, *Erskine Ramsay: His Life and Achievements* (New York: Cartwright & Ewing, 1942); Ramsay, "Some Personal and Family Notes," unpaginated, Ramsay papers. See also Owen, *Dictionary of Alabama Biography*, vol. 4, pp. 1406–8.

12. Armes, *Coal and Iron in Alabama*, p. 492.

13. Ibid., p. 494.

14. *Birmingham Age-Herald*, April 9, 1911.

15. Knight and Garth (or Gart) had been sentenced in Jefferson County but were from Dallas County. See *Selma Journal*, April 9, 1911.

16. *Birmingham Age-Herald*, April 9, 1911. Several wire service

accounts reported that during the afternoon trapped miners could be heard knocking on pipes. See *Savannah Morning News*, April 9, 1911. In fact, there was apparently no sign of life after the first miners escaped.

17. Emmet O'Neal to James G. Oakley, April 8, 1911, O'Neal papers, Alabama Department of Archives and History (hereafter cited as ADAH), Montgomery, Alabama.

18. Emmet O'Neal, *Convict Department, Its Management* (Montgomery: Brown Printing, 1913), pp. 3–4, 40–41.

19. *Birmingham Age-Herald*, April 10, 1911.

20. *Brewton Pine Belt News*, April 13, 1911; *Mobile Register*, April 11, 1911; *Macon Daily Telegraph*, April 11, 1911.

21. *Birmingham News*, April 8, 1911.

22. *Northport West Alabama Breeze*, April 19, 1911.

23. *Birmingham News*, April 8, 1911.

24. *Savannah Morning News*, April 10, 1911; *Macon Daily Telegraph*, April 10, 1911.

25. *Birmingham Ledger*, April 10, 1911; *Birmingham News*, April 10, 1911.

26. R. A. Statham to *Birmingham Labor Advocate*, April 28, 1911. See *Report of the Inspector of Alabama Coal Mines* for 1904, 1909, 1910, and 1914.

27. *Birmingham News*, April 10, 1911.

28. Charles J. Langston, writing in *Jasper Walker County News*, April 11, 1911.

29. *Birmingham News*, April 10, 1911; *Macon Daily Telegraph*, April 10, 1911; *Montgomery Advertiser*, April 12, 1911.

30. Langston, writing in *Jasper Walker County News*, April 11, 1911.

31. *Tuscaloosa News*, April 11, 1911.

32. Before leaving for Washington, Holmes stressed the thoroughness of the Banner search: "After previous disasters when men were found alive after days, we wished to take no chances. We went to every nook and corner of the mine to see that no one was alive." See Holmes's statement in *Savannah Morning News*, April 12, 1911, and *Macon Daily Telegraph*, April 12, 1911.

33. *Quadrennial Report, Board of Inspectors of Convicts, 1910–1914*, pp. 186–98.

34. *Indianapolis* [Indiana] *United Mine Workers Journal,* April 20, 1911, quoting Jeff Bailey in the *Birmingham Ledger.* Bailey insisted that farm life in the Black Belt was infinitely better than working in a mine; a mine's unfamiliar and dark interior alone was frightening to the convicts.

Chapter 2. The Convict Lease System

1. Governor John Murphy to the Senate and House of Representatives, November 21, 1826, Governor's Message, 1824–1827, ADAH.

2. William Schley to C. C. Clay, October 11, 1836; Robert Dunlop to C. C. Clay, September 16, 1836; C. C. Clay papers, ADAH; Malcolm Smith and Alvin A. McWhorter to Governor A. P. Bagby, February 23, 1839, A. P. Bagby papers, ADAH.

3. Elizabeth Porter, *A History of Wetumpka* (Wetumpka: Chamber of Commerce, 1957), pp. 27–28. For Alabama see Governor Rufus W. Cobb, "History of the Penitentiary," in *First Biennial Report of the Inspectors of Convicts to the Governor: From October 1, 1884, to October 1, 1886* (Montgomery: Barrett, 1886), p. 348; Jack Leonard Lerner, "A Monument to Shame: The Convict Lease System in Alabama" (Master's thesis, Samford University, 1969); Elizabeth Bonner Clark, "Abolition of the Convict Lease System in Alabama, 1913–1928" (Master's thesis, University of Alabama, 1949); R. H. Dawson, "The Convict System of Alabama, As It Was and As It Is," in *Hand-Book of Alabama: A Complete Index to the State, with Map,* ed. Safford Berney (Birmingham: Roberts & Son, 1892), pp. 254–66; Malcolm Moos, *State Penal Administration in Alabama* (University, Ala.: Bureau of Public Administration, 1942).

4. R. H. Dawson, "Convict System of Alabama," p. 256.

5. Ibid. See also Lerner, "Monument to Shame," p. 59, and Clark, "Abolition of the Convict Lease System," pp. 11–12.

6. *The Penal Code of Alabama Adopted by the General Assembly at Session, 1865–1866* (Montgomery: Reid & Screws, 1866), pp. 10–11. For the general picture see Theodore B. Wilson, *The Black Codes of the South* (University: University of Alabama Press, 1965).

7. *Penal Code, 1865–1866,* sec. 217–34. See also Blake F. Mc-

Kelvey, "Penal Slavery and Southern Reconstruction," *Journal of Negro History* 20 (April 1935), pp. 153–79.

8. Lerner, "Monument to Shame," p. 64.

9. Cobb, "History of the Penitentiary," p. 352; Horace Mann Bond, *Negro Education in Alabama: A Study in Cotton and Steel* (Washington, D.C.: Associated Publishers, 1939), p. 61n.

10. Clark, "Abolition of the Convict Lease System," pp. 13–14.

11. Jean E. Keith, "The Role of the Louisville and Nashville Railroad Company in the Early Development of Alabama Coal and Iron," *Bulletin of the Business Historical Society* 26 (September 1952), pp. 165–74. See also James F. Doster, *Railroads in Alabama Politics, 1875–1914* (University: University of Alabama Press, 1957). Several works on Birmingham have high merit. Armes, *Coal and Iron in Alabama*, was a pioneering study. Neither critical nor well balanced, it is nevertheless well written, factual, and thorough in its treatment of personalities and remains an invaluable source. The city's unique place in Alabama is deftly sketched in Virginia Van der Veer Hamilton, *Alabama: A Bicentennial History* (New York: Norton, 1977), pp. 127–48. Marjorie Longenecker White, *The Birmingham District: An Industrial History and Guide* (Birmingham: Birmingham Historical Society, 1981), is a useful survey. Two important studies combining much research and interpretation are Martha C. Mitchell, "Birmingham: Biography of a City of the New South" (Ph.D. diss., University of Chicago, 1946), and especially Carl V. Harris, *Political Power in Birmingham, 1871–1921* (Knoxville: University of Tennessee Press, 1977). See also Leah Rawls Atkins, *The Valley and the Hills: An Illustrated History of Birmingham and Jefferson County* (Woodland Hills, Calif.: Windsor Publications, 1981); John C. Henley, Jr., *This Is Birmingham: The Story of the Founding and Growth of an American City* (Birmingham: Southern University Press, 1960); and Blaine A. Brownell, "Birmingham, Alabama: New South City in the 1920s," *Journal of Southern History* 38 (February 1972), pp. 21–48.

12. Justin Fuller, "History of the Tennessee Coal, Iron, and Railroad Company, 1852–1907" (Ph.D. diss., University of North Carolina, 1966). For discord within the Bourbon coalition see Jonathan M. Wiener, *Social Origins of the New South: Alabama, 1860–1885* (Baton Rouge: Louisiana State University Press, 1978). The best

book on the Bourbons is Allen Johnston Going, *Bourbon Democracy in Alabama, 1874–1890* (University: University of Alabama Press, 1951).

13. *Report of the Joint Committee to Inspect the State Penitentiary and State Farm* (Montgomery: Screws, 1875), pp. 8–9. See also *Annual Report of the Inspectors of the Alabama Penitentiary from March 1st to September 30, 1873* (Montgomery: Bingham, 1873), and *Annual Report of the Inspectors of the Alabama Penitentiary from October 1, 1873, to September 30, 1874* (Montgomery: Screws, 1874).

14. *Report of the Joint Committee to Inspect the State Penitentiary from March 4, 1875* (Montgomery: Screws, 1876), p. 4; "Remoc B" in *Eufaula Times and News*, March 7, 1882. See also Virgil P. Koart, "The Administrations of George Smith Houston, Governor of Alabama, 1874–1878" (Master's thesis, Auburn University, 1963), pp. 134–36.

15. William O. Scoggs, "Convict and Apprenticed Labor in the South," in *Economic History 1865–1909*, vol. 6 of *The South in the Building of the Nation*, 12 vols., ed. Julian A. C. Chandler, et al. (Richmond: Southern Historical Publication Society, 1919), p. 48. For harsh contemporary criticism see George W. Cable, *The Silent South* (New York: Scribner's Sons, 1885). Cable also made his views well known in articles published in *Century Magazine*. For recent scholarship see Fletcher M. Green, "Some Aspects of the Convict Lease System in the Southern States," in *Essays in Southern History*, ed. Fletcher M. Green (Chapel Hill: University of North Carolina Press, 1949), pp. 112–23; Dan T. Carter, "Prisons, Politics, and Business: The Convict Lease System in the Post Civil War South" (Master's thesis, University of Wisconsin, 1964); Hilda Zimmerman, "Penal Systems and Penal Reforms in the South since the Civil War" (Ph.D. diss., University of North Carolina, 1947); Pete Daniel, *The Shadow of Slavery: Peonage in the South, 1901–1969* (Urbana: University of Illinois Press, 1972); Daniel A. Novak, *The Wheel of Servitude: Black Forced Labor after Slavery* (Lexington: University of Kentucky Press, 1978); and Matthew J. Mancini, "Race, Economics, and the Abandonment of Convict Leasing," *Journal of Negro History* 63 (Fall 1978), pp. 339–52. See also Edward L. Ayers, *Vengeance and Justice: Crime and Punishment in the Nineteenth Century Amer-*

ican South (New York: Oxford University Press, 1984), pp. 34–72; Dewey W. Grantham, *Southern Progressivism: The Reconciliation of Progress and Tradition* (Knoxville: University of Tennessee Press, 1983), pp. 127–38; and Joel Williamson, *The Crucible of Race: Black-White Relations in the American South since Emancipation* (New York: Oxford University Press, 1984).

16. *Annual Report of the Inspectors of the Alabama Penitentiary for the Year Ended September 30, 1877, to the Governor* (Montgomery: Barrett & Brown, 1878), pp. 4, 6–7.

17. See "Remoc B" in *Eufaula Times and News,* March 7, 1882; Cobb, "History of the Penitentiary," p. 361.

18. *Report of the Joint Committee to Inquire into the Treatment of Convicts Employed in the Mines, and on Convict Farms, and in Any Other Place in the State* (Montgomery: Allred & Beers, 1881), p. 4.

19. See Rufus W. Cobb papers, ADAH, for a folder overflowing with letters endorsing Bankhead.

20. *Alabama Official and Statistical Register, 1907* (Montgomery: Brown Printing, 1907), p. 404.

21. The John Hollis Bankhead papers, ADAH, provide insight into his career, but they contain nothing about his tenure as warden. Bankhead's views are available only in newspapers and official documents.

22. *Biennial Report of the Inspectors of the Alabama Penitentiary, from September 30, 1880, to September 30, 1882* (Montgomery: Allred & Beers, 1882), pp. 15, 12–13.

23. Cobb, "History of the Penitentiary," p. 364.

24. *Acts,* 1882–1883 (Montgomery: Brown, 1883), p. 134.

25. *Biennial Report of the Inspectors of the Alabama Penitentiary to the Governor, 1884* (Montgomery: Brown, 1884), pp. 69, 6, 131, 25; *Report of the Joint Committee of the General Assembly, Appointed to Examine into the Convict System of Alabama, Session of 1888–89* (Montgomery: Brown Printing, 1889), p. 24.

26. *Report of the Inspectors, 1884,* pp. 70–71, 73; *Acts,* 1882–1883, p. 138. The Reginald Heber Dawson Diaries, ADAH, are an invaluable source and poignantly and painfully reveal how Dawson learned his job. See, for example, the entry for June 11, 1883. On

October 25, 1883, Dawson completed his inspection and wrote, "Left here more than ever disgusted with mines."

27. *Report of the Inspectors, 1884,* pp. 71–72.

28. *Acts,* 1884–1885, pp. 187–88, 192, 195.

29. *Report of the Inspectors of Convicts to the Governor, from October 1, 1888, to September 30, 1890* (Montgomery: Brown, 1890), p. 23.

30. *Report of the Inspectors of Convicts to the Governor from October 1, 1890, to August 31, 1892* (Montgomery: Smith, Allred & Company, 1892), p. 20.

31. William Warren Rogers, *The One-Gallused Rebellion: Agrarianism in Alabama, 1865–1896* (Baton Rouge: Louisiana State University Press, 1970), pp. 53, 213.

32. *Acts,* 1888–1891, pp. 18–19.

33. *Report of the Commission "For the Improvement of the Penal and Convict System of Alabama," Organized under an Act Approved February 18th, 1891* (Montgomery: n.p., 1891), p. 4.

34. For message of Governor Thomas G. Jones, November 16, 1892, see *Alabama Senate Journal,* 1892–1893 (Montgomery: Roemer Printing, 1893), pp. 14–47, especially 26–29.

35. *An Act to Create a New Convict System for the State of Alabama and to Provide for the Government, Discipline, and Maintenance of All Convicts in the State of Alabama* (Montgomery: n.p., 1892–1893), p. 17.

36. Ibid., pp. 21, 24–29; Clark, "Abolition of Convict Leasing System," p. 18.

37. *First Biennial Report of the Board of Managers of Convicts, to the Governor from September 1, 1892, to August 31, 1894* (Montgomery: Brown Printing, 1894), pp. 7–10.

38. *Report of the Inspectors, 1896,* pp. 2–3.

39. Ibid., pp. 8–9, 26, 33, 43. For a discussion of death among black convicts in 1895 at a single coal mine in Alabama, see Robert David Ward and William Warren Rogers, "Racial Inferiority, Convict Labor, and Modern Medicine: A Note on the Coalburg Affair," *Alabama Historical Quarterly* 44 (Fall and Winter 1982), pp. 203–10.

40. *Report of Special Committee to Investigate the Convict System, 1897* (Montgomery: n.p., n.d.), p. 22.

41. Ibid.

42. For the 1896 campaign see David A. Harris, "Racists and Reformers: A Study of Progressivism in Alabama, 1896–1911" (Ph.D. diss., University of North Carolina, 1967); Julian C. Braswell, "Senator Joseph Forney Johnston: Brinksmanship and Progressivism" (Master's thesis, Auburn University, 1969); and Rogers, *One-Gallused Rebellion,* pp. 293–317.

43. Alabama Progressivism has received scholarly attention, with excellent results. See D. Harris, "Racists and Reformers"; Sheldon Hackney, *Populism to Progressivism in Alabama* (Princeton: Princeton University Press, 1969), and Allen W. Jones, "A History of the Direct Primary in Alabama, 1840–1903" (Ph.D. diss., University of Alabama, 1964). The definitive book on Alabama's constitutions is Malcolm Cook McMillan, *Constitutional Development in Alabama, 1798–1901: A Study in Politics, the Negro, and Sectionalism* (Chapel Hill: University of North Carolina Press, 1955). For the political power and importance of railroads during the period, see Doster, *Railroads in Alabama Politics*. In addition, Jones and Doster have published a number of articles on Progressivism in the *Alabama Review.*

44. For perceptive interpretations of Southern Progressivism, see C. Vann Woodward, *Origins of the New South, 1877–1913* (Baton Rouge: Louisiana State University Press, 1951), pp. 369–95, and George Brown Tindall, *The Emergence of the New South, 1913–1945* (Baton Rouge: Louisiana State University Press, 1967), pp. 1–32, 219–53. Indispensable to an understanding of the movement is Grantham, *Southern Progressivism*. See also his "Contours of Southern Progressivism," *American Historical Review* 86 (December 1981), pp. 1035–59.

45. Green, "Aspects of the Convict Lease System," p. 115; Ayers, *Vengeance and Justice,* p. 222.

46. *Third Biennial Report of the Board of Inspectors of Convicts to the Governor, from September 1, 1898, to August 31, 1900* (Montgomery: Roemer, 1900), pp. 6–7, 13.

47. *Report of the Joint Committee of the General Assembly of Alabama upon the Convict System of Alabama* (Montgomery: Brown Printing, 1901), pp. 18, 15, 21.

48. Ibid., p. 29. See also Lerner, "Monument to Shame," pp. 144–48. In 1900, Governor Johnston listed for the General Assembly the progress and accomplishments of his administration. He boasted, "The convict department, from being a drag on our treasury, is now contributing nearly $50,000 annually to lessen the burden of the people." See *Acts*, 1900–1901, p. 9.

49. *Fourth Biennial Report of the Board of Inspectors of Convicts to the Governor: From September 1, 1900, to August 31, 1902* (Montgomery: Brown Printing, 1902), pp. 7–9.

50. *Fifth Biennial Report of the Board of Inspectors of Convicts to the Governor: From September 1, 1902, to August 31, 1904* (Montgomery: Brown Printing, 1904), pp. 8, 13–14, 29.

51. *Acts*, Special Session, 1907, pp. 179–89; Lerner, "Monument to Shame," pp. 154–55; Hackney, *Populism to Progressivism*, pp. 265, 271, 302.

52. *Alabama Official and Statistical Register, 1913* (Montgomery: Brown Printing, 1913), p. 315.

53. Melton A. McLaurin, *The Knights of Labor in the South* (Westport, Conn.: Greenwood Press, 1978), pp. 56, 187; Sidney H. Kessler, "The Organization of Negroes in the Knights of Labor," *Journal of Negro History* 37 (July 1952), pp. 224–76; Frederick Meyers, "The Knights of Labor in the South," *Southern Economic Journal* 6 (April 1940), pp. 479–87; John Abernathy, "The Knights of Labor in Alabama" (Master's thesis, University of Alabama, 1960).

54. Going, *Bourbon Democracy*, pp. 170–90; Frances Roberts, "William Manning Lowe and the Greenback Party in Alabama," *Alabama Review* 5 (April 1952), pp. 100–121; Herbert G. Gutman, "Black Coal Miners and the Greenback-Labor Party in Redeemer Alabama, 1878–1879: The Letters of Warren D. Kelley, Willis Johnson, Thomas, 'Dawson,' and Others," *Labor History* 10 (Summer 1969), pp. 506–35; Holman Head, "The Development of the Labor Movement in Alabama prior to 1900" (Master's thesis, University of Alabama, 1955); Frank McDonald Duke, "The United Mine Workers of America in Alabama: Industrial Unionism and Reform Legislation, 1890–1911" (Master's thesis, Auburn University, 1979); Philip Taft, *Organizing Dixie: Alabama Workers in the Industrial Era*, rev. and ed. Gary M. Fink (Westport, Conn.: Greenwood Press, 1981), pp.

3–30; Chris Evans, *History of the United Mine Workers of America*, 2 vols. (n.p.: n.p., 1920); F. Ray Marshall, *Labor in the South* (Cambridge: Harvard University Press, 1967), pp. 73–74.

55. *Acts*, 1890–1891, pp. 1362–65.

56. For message of Governor Thomas G. Jones, November 14, 1894, see *Alabama House Journal*, 1894–1895 (Montgomery: Roemer Printing, 1895), p. 44; for the strike see Robert D. Ward and William W. Rogers, *Labor Revolt in Alabama: The Great Strike of 1894* (University: University of Alabama Press, 1965), and Joseph Harold Goldstein, "Labor Unrest in the Birmingham District, 1871–1894" (Master's thesis, University of Alabama, 1951).

57. Duke, "UMW in Alabama," pp. 48–61; John Mitchell, *Organized Labor, Its Problems, Purposes, and Ideals and the Present and Future of American Wage Earners* (Philadelphia: American Book and Bible House, 1903), p. 89; Paul B. Worthman, "Black Workers and Labor Unions in Birmingham, Alabama, 1897–1904," *Labor History* 10 (Summer 1969), pp. 374–407; M. Mitchell, "Birmingham: Biography of a City," pp. 117–41. See also Herbert R. Northrup, *Organized Labor and the Negro* (New York: Harper & Brothers, 1944), and Philip S. Foner, *Organized Labor and the Black Worker, 1619–1973* (New York: Praeger Publishers, 1974).

58. Hackney, *Populism to Progressivism*, pp. 255–87; Duke, "UMW in Alabama," pp. 72–81; C. Harris, *Political Power in Birmingham*, pp. 221–22.

59. See editorial written later in *Indianapolis United Mine Workers Journal*, April 20, 1911.

60. See three studies by Richard Alan Straw: " 'This Is Not a Strike, It Is Simply a Revolution': Birmingham Miners Struggle for Power, 1894–1904" (Ph.D. diss., University of Missouri-Columbia, 1980); "The Collapse of Biracial Unionism: The Alabama Coal Strike of 1908," *Alabama Historical Quarterly* 37 (Summer 1975), pp. 92–114; and "Soldiers and Miners in a Strike Zone: Birmingham, 1908," *Alabama Review* 38 (October 1985), pp. 289–308. See also Nancy R. Elmore, "The Birmingham Coal Strike of 1908" (Master's thesis, University of Alabama, 1966); J. Wayne Flynt, "Alabama White Protestantism and Labor, 1900–1914," *Alabama Review* 25 (July 1972), pp. 193–217; Duke, "UMW in Alabama," pp. 82–88; C. Harris, *Political Power in Birmingham*, pp. 217–23.

61. Quoted in C. Harris, *Political Power in Birmingham,* pp. 203–4. See also Shelby M. Harrison, "A Cash-Nexus for Crime," *Survey* 27 (October 1911–March 1912), p. 1547.

62. Langston, writing in *Jasper Walker County News,* April 11, 1911.

Chapter 3. "The Facts Should Be Known"

1. J. L. Clemo to O'Neal, April 19, 1911, O'Neal papers; *Birmingham News,* April 13, 1911; *Birmingham Labor Advocate,* April 14, 1911; *Indianapolis United Mine Workers Journal,* March 16, 1911.

2. *Indianapolis United Mine Workers Journal,* April 13, 1911; Hugh Lynch to O'Neal, June 19, 1911, O'Neal papers; "Arkansas Traveler" to *Indianapolis United Mine Workers Journal,* March 16, 1911.

3. Duke, "UMW in Alabama," p. 98; *Birmingham News,* April 10, 1911; *Mobile Register,* April 12, 1911.

4. *Birmingham Howle's Iconoclast,* April 15, 1911.

5. *St. Louis Post-Dispatch,* April 11, 1911. See also *Jacksonville Times-Union,* April 9–12, 1911; *Atlanta Constitution,* April 9–10, 1911; *Washington Post,* April 9–10, 1911; *Times* (London), April 10, 1911; and *San Francisco Chronicle,* April 8, 1911.

6. *Anniston Evening Star,* April 13, 1911, quoting *Bessemer Standard.* Birmingham's black newspapers were *Hot Shots, Wide-Awake,* and *Christian Hope.* See Allen Woodrow Jones, "Alabama," in *The Black Press in the South, 1865–1979,* ed. Henry Lewis Suggs (Westport, Conn.: Greenwood Press, 1983), pp. 23–64.

7. E. Stagg Whitin, *Penal Servitude* (New York: National Committee on Prison Labor, 1912), p. 1.

8. *Tuscaloosa News,* April 12, 1911; see also issue of April 11, 1911. *Jasper Walker County News,* April 11, 1911; see also Harrison, "Cash-Nexus for Crime," p. 1539.

9. *Birmingham Labor Advocate,* April 14, 1911; *Indianapolis United Mine Workers Journal,* April 13, 1911.

10. *Talladega Reporter,* April 15, 1911; *Montgomery Journal,* April 18, 1911, quoting *Jackson South Alabamian; Birmingham Howle's*

Iconoclast, April 22, 1911; *Jasper Walker County News*, April 11, 1911.

11. *Montgomery Times*, April 11, 1911; *Atmore Spectrum*, April 13, 1911; *Brewton Pine Belt News*, April 13, 1911.

12. *Bay Minette Baldwin Times*, April 13, 1912.

13. *Anniston Weekly Times*, April 13, 1911; *Marion Standard*, April 14, 1911.

14. *Birmingham News*, April 10, 1911; *Birmingham Labor Advocate*, April 14, 1911; *Atmore Spectrum*, April 13, 1911; *Marion Standard*, April 14, 1911.

15. *Montgomery Advertiser*, April 12, 1911; *Birmingham News*, April 10, 1912; *Brewton Pine Belt News*, April 13, 1911.

16. *Birmingham News*, April 14, 1911.

17. *Birmingham Ledger*, April 10, 1911; *Montgomery Journal*, April 14, 1911; *Birmingham Ledger*, April 10, 1911; *Birmingham Age-Herald*, April 9, 1911; *Jasper Walker County News*, April 11, 1911.

18. *Montgomery Advertiser*, April 21, 1911; *Birmingham Times*, May 5, 1911. See also Anne Gary Pannell and Dorothea E. Wyatt, *Julia S. Tutwiler and Social Progress in Alabama* (University: University of Alabama Press, 1961), pp. 106–18.

19. John de B. Hooper to O'Neal, April 10, 1911, O'Neal papers.

20. Quoted in *Anniston Weekly Times*, April 13, 1911; see also Grace Hooten Gates, *The Model City of the New South: Anniston, Alabama, 1872–1900* (Huntsville, Ala.: Strode Publishers, 1978), pp. 232–33; Langston, writing in *Jasper Walker County News*, April 11, 1911.

21. For a statement of Whitin's philosophy see his letter to *New York Times*, April 15, 1911. For McNeel see *Montgomery Advertiser*, April 8, 1911.

22. Whitin to O'Neal, May 1, 1911, O'Neal papers; *Gadsden Evening Journal*, April 12, 1911.

23. James H. Strickland, director of public relations for Alabama By-Products, to William W. Rogers, November 14, 1984. Alabama By-Products absorbed Pratt Consolidated in 1925. The authors are grateful to Mr. Strickland for his efforts to locate the company records.

24. *Linden Democrat-Reporter,* April 13, 1911; *Birmingham Age-Herald,* April 10, 1911; *Gadsden Evening Journal,* April 11, 1911.
25. O'Neal to Whitin, April 25, 1911, O'Neal papers.
26. Whitin to O'Neal, May 1, 1911, O'Neal papers.
27. Ibid.
28. J. D. McNeel to Whitin, May 13, 1911, O'Neal papers.
29. John E. Milholland to Joseph Manning, April 10, 1911, O'Neal papers. For a sketch of the fifty-one-year-old Milholland, see *Who's Who in America,* 6th ed., s.v. "Milholland, John."
30. Pardons Book, State of Alabama, A–Z, May 15, 1911, p. 260, ADAH; see also *Birmingham News,* May 10, 1911, and *Birmingham Age-Herald,* May 15, 1911.
31. *Birmingham News,* April 11, 1911.
32. *Birmingham Ledger,* April 10, 1911. In a commentary on the difficulties facing a black worker, one of the prisoners said that, considering housing and food plus what he could earn by working overtime, he made more in prison than on the free market. The prisoner hastened to add, "if the sentence isn't too long." See *Birmingham News,* April 13, 1911.
33. *New York Times,* April 9, 1911; *New York Tribune,* April 9, 1911; *Birmingham News,* April 14, 1911, quoting *New York American;* Frederick W. Horton, comp., *Coal-Mine Accidents in the United States, 1896–1912* (Washington, D.C.: Government Printing Office, 1913), p. 70. Mitchell was quoted in the *Washington Post,* April 9, 1911; see also William Graebner, *Coal-Mining Safety in the Progressive Period: The Political Economy of Reform* (Lexington: University of Kentucky Press, 1976), p. 134.
34. See Inez Norton Edwards, "Emmet O'Neal: Alabama Governor, 1911–1915" (Master's thesis, Auburn University, 1957).
35. *Montgomery Times,* April 11, 1911.

Chapter 4. "At the Mercy of the Earth": The Mine Safety Law

1. Graebner, *Coal-Mining Safety,* p. 1. See also pp. 12–42 and his "Great Expectations: The Search for Order in Bituminous Coal, 1890–1917," *Business History Review* 48 (Spring 1974), 47–72.

2. Horton, *Coal-Mine Accidents,* pp. 18–26; *Birmingham News,* May 7, 1910; April 8 and 10, 1911; January 2, 1912; *Montgomery Advertiser,* May 17, 1910; *Indianapolis United Mine Workers Journal,* April 13, 1911.

3. Horton, *Coal-Mine Accidents,* pp. 26, 38; *Birmingham News,* April 10, 1911; Albert H. Fay, comp., *Metal-Mine Accidents during the Calendar Year 1911* (Washington, D.C.: Government Printing Office, 1913), p. 16.

4. *Montgomery Advertiser,* May 18, 1910.

5. *Birmingham News,* May 17, 1910.

6. Graebner, *Coal-Mining Safety,* p. 2.

7. *Birmingham Age-Herald,* April 13, 1911, quoting *New York Herald; Alabama House Journal,* 1911, vol. 1, pp. 260–61.

8. *Alabama Senate Journal,* 1911, vol. 1, p. 411; *Alabama Official and Statistical Register, 1911* (Montgomery: Brown Printing, 1912), p. 80; *Alabama Senate Journal,* 1911, vol. 1, p. 545, and vol. 2, p. 2021.

9. *Mobile Register,* April 12, 1911; *Alabama House Journal,* 1911, vol. 1, p. 724.

10. *Alabama Official and Statistical Register, 1911,* p. 134; *Atmore Spectrum,* April 13, 1911. For a direct accusation of collusion between Hollis and the Coal Operators' Association of Alabama, see *Birmingham Labor Advocate,* February 10, 1911.

11. *Alabama House Journal,* 1911, vol. 1, p. 1222.

12. *Birmingham Labor Advocate,* February 3, 1911. The letter was reprinted with favorable comment in *Indianapolis United Mine Workers Journal,* February 16, 1911.

13. *Birmingham Labor Advocate,* February 3, 1911; *Indianapolis United Mine Workers Journal,* February 16, 1911.

14. *Indianapolis United Mine Workers Journal,* February 16, 1911. For an early denunciation of the bill see *Birmingham Labor Advocate,* January 27, 1911.

15. *Alabama Official and Statistical Register, 1911,* p. 117; *Alabama House Journal,* 1911, vol. 2, pp. 2342–76. The UMW's officials and others sympathetic to the miners helped draft the O'Neill bill. See *Birmingham Labor Advocate,* March 17 and April 7, 1911.

16. *Atmore Spectrum,* April 13, 1911 (see also *Birmingham Age-Herald,* April 20, 1911); *Birmingham News,* April 12, 1911; *Birming-*

ham Age-Herald, April 20, 1911; *Alabama House Journal,* 1911, vol. 2, pp. 2377–82.

17. *Alabama House Journal,* 1911, vol. 2, p. 2431; *Montgomery Journal,* April 10, 1911; *Montgomery Advertiser,* April 11, 1911; *Alabama Official and Statistical Register, 1911,* pp. 33, 115.

18. See *Montgomery Journal,* April 10, 1911; *Birmingham Ledger,* April 10, 1911; *Marion Standard,* April 14, 1911; and *Brewton Pine Belt News,* April 13, 1911.

19. John de B. Hooper to O'Neal, April 10, 1911, and John M. Russell to O'Neal, April 12, 1911, O'Neal papers.

20. R. A. Statham to *Indianapolis United Mine Workers Journal,* March 9, 1911.

21. *Montgomery Advertiser,* April 11, 1911; *Birmingham Ledger,* April 11, 1911. For a similar view see *Birmingham News,* April 10, 1911.

22. *Alabama Senate Journal,* 1911, vol. 2, pp. 1953, 2012.

23. *Birmingham News,* April 12, 1911. For convenience, see *Mining Law of the State of Alabama* (Montgomery: Brown Printing, 1911), pp. 1–48. *Alabama Senate Journal,* 1911, vol. 1, pp. 2133–74, 2175, 2201.

24. *Birmingham News,* April 12, 1911; *Alabama House Journal,* 1911, vol. 2, pp. 2658–99; *Alabama Official and Statistical Register, 1911,* p. 112; *Montgomery Advertiser,* April 12, 1911; *Birmingham Age-Herald,* April 13, 1911.

25. *Birmingham News,* April 11, 1911; *Montgomery Journal,* April 12, 1911; *Mobile Register,* April 12, 1911.

26. O'Neal, *Convict Department,* p. 113; R. A. Statham, quoted in *Indianapolis United Mine Workers Journal,* March 16, 1911. See also Harry Bonefield to *Birmingham Labor Advocate,* March 17, 1911. See O'Neill's remarks in *Birmingham News,* April 17, 1911.

27. *Birmingham Age-Herald,* April 16 and 20, 1911; *Birmingham News,* April 26, 1911; *Talladega Reporter,* April 15, 1911; *Acts,* 1911, pp. 500–543. See also *Montgomery Advertiser,* April 15, 1911; *Birmingham Labor Advocate,* April 21 and 28, 1911; and Duke, "UMW in Alabama," p. 101.

28. O'Neal, *Convict Department,* pp. 113–14.

29. *Indianapolis United Mine Workers Journal,* April 13, 1911; *Birmingham News,* April 17, 1911.

30. *Mining Law of Alabama,* 1911.
31. *Birmingham Journal,* July 27, 1911.

Chapter 5. "A Principle of Justice": The Convict Lease Question

1. For Alabama's involvement in peonage see Daniel, *Shadow of Slavery,* pp. 43–64, 65–81. See also William Cohen, "Negro Involuntary Servitude in the South, 1865–1940: A Preliminary Analysis," *Journal of Southern History* 42 (February 1976), pp. 31–60. For the convict lease system in other states see Jesse Crawford Crowe, "Agitation for Penal Reform in Tennessee, 1870–1900" (Ph.D. diss., Vanderbilt University, 1954); Randall G. Shelden, "From Slave to Caste Society: Penal Changes in Tennessee, 1830–1915," *Tennessee Historical Quarterly* 38 (Winter 1979), pp. 462–78; Mark T. Carleton, *Politics and Punishment: The History of the Louisiana State Penal System* (Baton Rouge: Louisiana State University Press, 1971); A. Elizabeth Taylor, "The Origin and Development of the Convict Lease System in Georgia," *Georgia Historical Quarterly* 16 (June 1942), pp. 113–38, and her "Abolition of the Convict Lease System in Georgia," ibid. (September–December 1942), pp. 273–87; Dewey Grantham, *Southern Progressivism,* p. 127; Jane Zimmerman, "The Convict Lease System in Arkansas and the Fight for Abolition," *Arkansas Historical Quarterly* 8 (Autumn 1949), pp. 171–88, and her "Penal Reform Movement in the South during the Progressive Era, 1890–1917," *Journal of Southern History* 17 (November 1951), pp. 462–92.

2. *Alabama Senate Journal,* 1911, vol. 2, p. 223.

3. Ibid., vol. 1, p. 223. About Moulthrop see Eugenia Persons Smartt, *History of Eufaula, Alabama* (Birmingham: Roberts & Son, 1933), pp. 271–72, and Mattie Thomas Thompson, *History of Barbour County, Alabama* (Eufaula, Ala.: published by the author, 1939), pp. 450–52.

4. Moulthrop writing to the *Mobile Register,* April 9, 1911. See also Duke, "UMW in Alabama," p. 100.

5. *Alabama Official and Statistical Register, 1911,* pp. 83–84.

6. *Alabama Senate Journal*, 1911, vol. 1, pp. 374, 470; *Alabama Official and Statistical Register, 1911,* pp. 91–92.

7. *Alabama Official and Statistical Register, 1911,* p. 74.

8. *Alabama Senate Journal*, 1911, vol. 1, pp. 503, 700–701, 793, 853–54.

9. *Alabama House Journal*, 1911, vol. 1, pp. 1180, 1183, 1500–1501; *Alabama Official and Statistical Register, 1911,* p. 101.

10. *Alabama House Journal*, 1911, vol. 1, pp. 179–80.

11. Ibid., pp. 738–39, 1379–88.

12. *Alabama Official and Statistical Record, 1911,* p. 121; *Alabama House Journal*, 1911, vol. 1, p. 1388.

13. *Alabama Official and Statistical Record, 1911,* p. 121; *Alabama House Journal*, 1911, vol. 1, p. 1391.

14. *Birmingham Times*, April 21, 1911 (for a similar editorial see *Birmingham Howle's Iconoclast*, April 22, 1911); quoted in *Montgomery Journal*, April 8, 1911; *Montgomery Journal*, April 18, 1911, quoting *Jackson South Alabamian*.

15. *Alabama Official and Statistical Record, 1911,* p. 115.

16. *Alabama House Journal*, 1911, vol. 2, p. 2469; *Birmingham News*, April 11, 1911; *Birmingham Labor Advocate*, April 14, 1911; *Birmingham Times*, April 14, 1911.

17. See *Atmore Spectrum*, April 13, 1911; *Mobile Register,* April 17, 1911.

18. *Northport West Alabama Breeze*, April 15, 1911; *Selma Journal*, April 14, 1911.

19. *Montgomery Journal*, April 18, 1911; *Montgomery Advertiser,* April 11, 1911.

20. *Montgomery Advertiser,* April 21, 1911.

21. *Birmingham Ledger,* April 19, 1911.

22. Undated newspaper clipping, Ward Scrapbook, vol. 10, 1911–1913, Birmingham Public Library. This scrapbook contains a clipping from *Birmingham Age-Herald*, January 19, 1912, showing that Maben had not changed his views. George B. Ward was a prominent, progressive civic leader and politician in Birmingham.

23. *Mobile Register,* April 9, 1911.

24. *Montgomery Journal*, April 14, 1911, quoting *Tuscaloosa News; Birmingham Howle's Iconoclast*, April 22, 1911; *Jasper Walker County News*, April 11, 1911.

25. *Birmingham News,* April 10, 1911. See *Birmingham Labor Advocate,* January 13 and 20, February 10, March 17, April 4 and 14, and May 19, 1911.

26. For a contemporary view see J. H. Jones, "Penitentiary Reform in Mississippi," *Publications of the Mississippi Historical Society* 6 (1902), pp. 111–28. See also Lyda Gordon Shivers, "A History of the Mississippi Penitentiary" (Master's thesis, University of Mississippi, 1930). Three incisive discussions by historians are William F. Holmes, *The White Chief: James Kimble Vardaman* (Baton Rouge: Louisiana State University Press, 1970), pp. 150–67; Albert D. Kirwan, *Revolt of the Rednecks* (Lexington: University of Kentucky Press, 1951), pp. 168–74; and Vernon Lane Wharton, *The Negro in Mississippi, 1865–1890* (Chapel Hill: University of North Carolina Press, 1947), pp. 234–42.

27. *Montgomery Advertiser,* April 21, 1911, quoting *Talladega Our Mountain Home; Birmingham Howle's Iconoclast,* April 22, 1911.

28. For convenience, the various citations in the discussion on the caucus are listed in one footnote.

29. For the caucus see *Birmingham Age-Herald,* April 12, 1911; *Mobile Register,* April 17, 1911; *Birmingham News,* April 13, 1911; *Montgomery Advertiser,* April 12, 1911; *Birmingham Ledger,* April 13, 1911. See *Alabama Official and Statistical Register, 1911* for sketches of the following men: William O. Mulkey, p. 116; Judge S. Williams, pp. 100–101; Frank Stollenwerck, Jr., p. 125; Peter B. Mastin, pp. 124–25; Fleetwood Rice, p. 133; John W. Green, p. 109; Thomas H. Molton, p. 116; Joseph H. James, Jr., p. 126; John H. Cranford, p. 134; James E. Jenkins, p. 104; and Josiah J. Arnold, pp. 102–3. For additional information on Molton see C. Harris, *Political Power in Birmingham,* pp. 211, 232–33, 238.

30. *Montgomery Advertiser,* April 14, 1911; *Birmingham Ledger,* April 13, 1911.

31. *Montgomery Journal,* April 15, 1911; *Birmingham Age-Herald,* April 15, 1911.

32. *Montgomery Advertiser,* April 15, 1911.

33. O'Neal to E. Stagg Whitin, April 25, 1911, O'Neal papers.

34. *Birmingham News,* May 22, 1911. Moulthrop's remarks were made in a speech to a national congress on good roads held at Birmingham May 23–26, 1911.

35. *Montgomery Advertiser,* April 15, 1911.

Notes to Chapter 6

Chapter 6. "According to the Best . . . Experts"

1. J. V. Allen to O'Neal, April 27, 1911, and J. D. McNeal to Charles H. Nesbitt, May 6, 1911, O'Neal papers.

2. The April 10, 1911, issues of the *Birmingham Ledger, Birmingham Age Herald,* and *Birmingham News*. All of these papers added editorial support to McCormack's statement.

3. Ramsay's article was reprinted in its entirety in the *Birmingham Age-Herald,* April 10, 1911.

4. Minutes, Board of Revenue, Jefferson County, March 1, 1911–August 8, 1911, p. 106. The minutes are filed at the Jefferson County Courthouse, Birmingham.

5. C. Harris, *Political Power in Birmingham,* pp. 203–6. Even a cursory look at the minutes of the Board of Revenue reveals how lucrative leasing was for Birmingham.

6. *Birmingham News,* April 13, 1911; *Birmingham Age-Herald,* April 11, 1911; *Birmingham Ledger,* April 17, 1911; *Tuscaloosa News,* April 14, 1911.

7. *Tuscaloosa News,* April 14, 1911.

8. The official records of the coroner's jury have been destroyed, but see extended reports in *Birmingham News,* April 13–16, 1911; *Montgomery Advertiser,* April 14 and 16, 1911; and *Birmingham Ledger,* April 7, 1911.

9. R. A. Statham, writing to the *Birmingham Labor Advocate,* April 28, 1911.

10. *Birmingham Age-Herald,* April 12, 1911.

11. *Montgomery Advertiser,* April 10, 1911, and *Birmingham Ledger,* April 11, 1911; Graebner, *Coal-Mining Safety,* p. 4.

12. *Birmingham News,* April 16, 1911; *Birmingham Ledger,* April 16 and 21, 1911. Rice's work in the bureau's experimental mine led to the discovery of the efficacy of rock dusting. See his "Explosibility of Coal Dust," *U.S. Bureau of Mines, Bulletin 20* (Washington, D.C.: Government Printing Office, 1911), and his later "Coal Dust Explosion Tests in the Experimental Mine, 1913 to 1918 Inclusive," *U.S. Bureau of Mines, Bulletin 167* (Washington, D.C.: Government Printing Office, 1922). And see J. J. Rutledge and Clarence Hall, "The Use of Permissible Explosives," *U.S. Bureau of Mines, Bulletin 10* (Washington, D.C.: Government Printing Office, 1912).

13. See letter of Charles Tinlin to *Indianapolis United Mine Workers Journal*, July 20, 1911.

14. *Montgomery Advertiser*, April 14 and 23, 1911; *Birmingham News*, April 13 and 16, 1911.

15. *Birmingham Age-Herald*, April 20, 1911.

16. Neil's and Dickinson's official reports are not on file at the ADAH. The full Neil report was reconstructed from the *Birmingham Age-Herald*, April 23, 1911, and from the *Birmingham News*, April 23, 1911. The fragmentary Dickinson report was assembled from several newspapers.

17. *Birmingham News*, April 23, 1911.

18. Ibid.

19. Ibid.

20. *Birmingham Age-Herald*, April 23, 1911.

21. *Montgomery Journal*, April 22, 1911; see also *Montgomery Advertiser*, April 23, 1911, and *Birmingham News*, April 22, 1911; *Birmingham Ledger*, April 22, 1911.

22. John M. Russell to O'Neal, April 12, 1911, O'Neal papers.

23. Rutledge's findings were quoted by H. B. Humphrey, "Historical Summary of Coal-Mine Explosions in the United States, 1810–1958," *Bureau of Mines, Bulletin 586* (Washington, D.C.: Government Printing Office, 1960), p. 49. See also F. C. Cash and H. B. Humphrey, "Explosions in Alabama Mines," *Bureau of Mines, Information Circular* 6352 (Washington, D.C.: Government Printing Office, 1930), pp. 1–8. For other important information from the Bureau of Mines, see Correspondence and Other Records Relating to Mining Operations, File 431, Box no. 2, Record Group 70, Federal Records Center, Suitland, Maryland.

24. *Birmingham Age-Herald*, April 23, 1911. See O'Neal to James E. Strong, April 6, 1911, O'Neal papers. Strong, General Superintendent of Coal Mines for Southern Iron and Steel Company of Birmingham, strongly supported Hillhouse. See also *Montgomery Journal*, April 21, 1911, and *Birmingham News*, April 24, 1911.

25. Jesse F. Stallings, telegram to O'Neal, April 22, 1911, Jesse F. Stallings papers, Southern Historical Collection, University of North Carolina, Chapel Hill.

26. *Birmingham News*, April 22, 1911.

27. *Montgomery Advertiser*, April 22, 1911; *Birmingham Age-Herald*, April 21–22, 1911; *Montgomery Journal*, April 22, 1911.

28. *Birmingham News,* May 22, 1911.

29. Ibid., May 2 and 15, 1911.

30. *Rules and Regulations for the Government of the Convicts of Alabama Adopted by the Board of Inspectors of Convicts, August 20, 1912* (Montgomery: Brown Printing, 1912).

31. *Birmingham News,* April 22 and May 10, 1911.

32. Final Record, Civil Cases, Circuit Court Jefferson County, 1911, vol. 93, pp. 231–32, 234–35. On file at the Jefferson County Courthouse, Birmingham. In two instances no disposition or settlement whatever was given for suits on behalf of deceased black prisoners. See ibid., pp. 304–5, 338.

33. Final Record, City Court [Birmingham], 1911–1912, vol. 201, pp. 1–4; vol. 203, pp. 191–92; vol. 206, pp. 469–70. On file at the Jefferson County Courthouse. See *Birmingham News,* June 22 and 28, 1911. In reporting cases, newspapers frequently confused the first names of deceased miners. Although not without error, court records were more reliable.

34. Final Record, City Court, 1911–1912, vol. 20? (last number missing), pp. 53–55, for Bowers case; pp. 47–49, for Brown case. See also Final Record, Circuit Court, 1911, vol. 91, pp. 287–88.

35. See Civil Docket of United States Circuit Court Northern District, Southern Division, books 2–5; and cases numbered between 1699 and 1847, Record Group 21, Federal Records Center, East Point, Georgia.

36. Minutes, Board of Revenue, Jefferson County, March 1, 1911–August 8, 1911, p. 265.

37. *Indianapolis United Mine Workers Journal,* August 3, 1911; see also *Birmingham News,* July 7–8, 1911.

38. *Birmingham News,* May 9, 1911.

Chapter 7. "The Moral Question Obtrudes"

1. O'Neal, *Convict Department,* p. 62.

2. Jesse F. Stallings to O'Neal, August–December 1911, Stallings papers.

3. A copy of the contract is located in the O'Neal papers. See also *Indianapolis United Mine Workers Journal,* September 7, 1911.

4. *Indianapolis United Mine Workers Journal*, September 7, 1911. See also *Birmingham News*, August 9 and 17, 1911.

5. Marlene Hunt Rikard, "An Experiment in Welfare Capitalism: The Health Care Services of the Tennessee Coal, Iron, and Railroad Company" (Ph.D. diss., University of Alabama, 1983), pp. 2–4. See also Rikard's "George Gordon Crawford: Man of the New South" (Master's thesis, University of Alabama, 1971); Fuller, "History of TCI," pp. 121–67; and W. G. Moore, "The Acquisition of the Tennessee Coal Company by the United States Steel Corporation in 1907" (Master's thesis, Samford University, 1971).

6. O'Neal, *Convict Department*, p. 51.

7. *Birmingham News*, January 1, 1912. See also issues of December 29–30, 1911.

8. *Quadrennial Report, Board of Inspectors of Convicts, 1910–1914*, pp. 54, 56.

9. *Birmingham News*, January 1, 1912.

10. Death Record Ledger Book, State Convict Bureau of Alabama, September 1, 1910, to August 31, 1915, unpaginated, ADAH; *Birmingham News*, December 7, 1912.

11. C. Harris, *Political Power in Birmingham*, pp. 206–7.

12. Edwards, "Emmet O'Neal," pp. 78–98; Clark, "Abolition of the Convict Lease System," pp. 38–50; O'Neal, *Convict Department*, pp. 64–101.

13. Ibid.

14. *Birmingham News*, May 22, 1911. Moulthrop never served in the Alabama legislature again.

15. For the name change see *Acts*, 1919, pp. 117–22; for ending the lease see pp. 522–23; for extending the lease see *Acts*, Special Session, 1921, pp. 30–31.

16. "After Florida, Alabama," *Nation* 92 (June 11, 1923), pp. 31–32; "Alabama's Convict System under Fire," *Literary Digest* 89 (April 10, 1926), pp. 10–11; Frank Tannenbaum, "Southern Prisons," *Century* 106 (June 1923), pp. 387–98. For the situation in Florida see N. Gordon Carper, "Martin Tabert, Martyr of an Era," *Florida Historical Quarterly* 52 (October 1973), pp. 115–31, and Tindall, *Emergence of the New South*, pp. 213–14.

17. *Acts*, 1927, pp. 51, 5–55; see also *Acts*, 1923, pp. 6–7, and "The End of Convict Leasing in Alabama," *Literary Digest* 98 (July 21, 1928), p. 11.

Notes to Chapter 7

18. The Alabama By-Products Corporation was formed in 1920 to mine coal and make coke. Its founders were Alabama businessmen Horace Hammond, Albert P. Bush, and Morris Bush. See White, *The Birmingham District*, p. 237.

19. *Quadrennial Report, Board of Inspectors of Convicts, 1910–1914*, pp. 41–42.

Select Bibliography

All of the sources for this study are cited in the notes. The materials mentioned here, while not comprehensive, contain some of the important works.

By far the most relevant manuscript collection was that of Governor Emmet O'Neal, located at the Alabama Department of Archives and History, Montgomery. Also of value at the ADAH were the diaries of an experienced mine inspector, Reginald Heber Dawson. The department's collection of John Hollis Bankhead papers proved disappointing for the period when he was warden of the Alabama prison system. The papers of Jesse Stallings, a Progressive Alabama politician, located at the Southern Historical Collection, University of North Carolina, Chapel Hill, proved useful. The views of management were well reflected in the Erskine Ramsay papers, Department of Archives and Manuscripts, Birmingham Public Library.

A wide variety of county, city, state, and national official records, both published and unpublished, were essential to the writing of this book. The city and circuit court records of Birmingham and of Jefferson County, respectively (Jefferson County Courthouse, Birmingham) and the records of the United States Circuit Court Northern District, Southern Division (Federal Records Center, East Point, Georgia) disclosed the disposition of various cases growing out of the tragedy at the Banner Mine. Information relative to the Banner explosion from the United States Bureau of Mines is housed in the Department of the Interior Library, Washington, D.C., and in the Federal Records Center, Suitland, Maryland.

Printed materials that were used extensively include the *Acts of the General Assembly of Alabama* (which we cite as *Acts*), *Journal of the Senate of the State of Alabama* (cited as *Alabama Senate Journal*), and *Journal of the House of Representatives of the State of Alabama* (cited as *Alabama House Journal*). Equally revealing were the extensive reports issued by inspectors, wardens, and officials of the prison system. Typical of such materials are *Rules and Regulations for the Government of the Convicts of Alabama Adopted by the Board of Inspectors of Convicts, August 20, 1912* (Montgomery: Brown Printing, 1912); *Report of the Joint Committee of the General Assembly of Alabama upon the Convict System of Alabama* (Montgomery: Brown Printing, 1901); and *Quadrennial Report of the Board of Inspectors of Convicts for the State of Alabama to the Governor, September 1, 1910–August 31st, 1914* (Montgomery: Brown Printing, 1915). All of these publications are located at the ADAH.

Alabama newspapers, besides reporting the events of the accident, disclosed the opinions of editors, of individuals, and of pressure groups and special interests. Voices sympathetic to Pratt Consolidated and the corporate view included county weeklies, but the most powerful advocates were several Birmingham newspapers: the *News*, *Age-Herald*, and *Ledger*. A similar position was taken by the *Montgomery Advertiser*. The *Mobile Register* was the single most important urban newspaper of general readership that pushed for reform and for ending the convict lease system. Defending the cause of labor and attacking the lease system were the *Birmingham Labor Advocate*, *Indianapolis* [Indiana] *United Mine Workers Journal*, and *Birmingham Howle's Iconoclast*.

Numerous articles in scholarly journals (see notes) and books on Alabama, the South, and the nation compose the secondary sources. Volumes useful for the state include Sheldon Hackney, *Populism to Progressivism in Alabama* (Princeton: Princeton University Press, 1969); Malcolm Cook McMillan, *Constitutional Development in Alabama, 1798–1901* (Chapel Hill: University of North Carolina Press, 1955); Allen Johnston Going, *Bourbon Democracy in Alabama, 1874–1890* (University: University of Alabama Press, 1951); James F. Doster, *Railroads in Alabama Politics, 1875–1914* (University: University of Alabama Press, 1957); Ethel M. Armes, *The Story of Coal and Iron in Alabama* (Birmingham: Chamber of Commerce, 1910); Robert David Ward and William Warren Rogers, *Labor Revolt in Alabama: The Great Strike of 1894* (University: University of Alabama Press, 1965); Marjorie Longenecker White, *The Birmingham District: An Industrial History and Guide* (Birmingham: Birmingham Historical Society, 1981); Horace Mann Bond, *Negro Education in Alabama: A Study in Cotton and Steel* (Washington, D.C.: Associated Publishers, 1939); and Carl V. Harris, *Political Power in Birmingham, 1871–1921* (Knoxville: University of Tennessee Press, 1977).

The time, setting, and both general and specific information are offered in several broader studies: Pete Daniel, *The Shadow of Slavery: Peonage in the South, 1901–1969* (Urbana: University of Illinois Press, 1972); Daniel A. Novak, *The Wheel of Servitude: Black Forced Labor after Slavery* (Lexington: University of Kentucky Press, 1978); Dewey W. Grantham, *Southern Progressivism: The Reconciliation of Progress and Tradition* (Knoxville: University of Tennessee Press, 1983); Edward L. Ayers, *Vengeance and Justice: Crime and Punishment in the Nineteenth Century American South* (New York: Oxford University Press, 1984); William Graebner, *Coal-Mining Safety in the Progressive Period: The Political Economy of Reform* (Lexington: University of Kentucky Press, 1976); C. Vann Woodward, *Origins of the New South, 1877–1913,* vol. 9 of *History of the South* (Baton Rouge: Louisiana State University Press, 1951); and George Brown Tindall, *The Emergence of the New South, 1913–1945,* vol. 10 of *History of the South* (Baton Rouge: Louisiana State University Press, 1967).

This book has benefited greatly from many excellent dissertations and theses. A partial listing follows: Jack Leonard Lerner, "A Monument to Shame: The Convict Lease System in Alabama" (Master's thesis, Samford University, 1969); Elizabeth Bonner Clark, "Abolition of the Convict Lease System in Alabama, 1913–1928" (Master's thesis, University of Alabama, 1949); Martha C. Mitchell, "Birmingham: Biography of a City of the New South" (Ph.D. diss., University of Chicago, 1946); Justin Fuller, "History of the Tennessee Coal, Iron, and Railroad Company, 1852–1907" (Ph.D. diss., University of North Carolina, 1966); Hilda Zimmerman, "Penal Systems and Penal Reforms in the South since the Civil War" (Ph.D. diss., University of North Carolina, 1947); David A. Harris, "Racists and Reformers: A Study of Progressivism in Alabama, 1896–1911" (Ph.D.

diss., University of North Carolina, 1967); Julian C. Braswell, "Senator Joseph Forney Johnston: Brinksmanship and Progressivism" (Master's thesis, Auburn University, 1969); Allen W. Jones, "A History of the Direct Primary in Alabama, 1840–1903" (Ph.D. diss., University of Alabama, 1964); Frank McDonald Duke, "The United Mine Workers of America in Alabama: Industrial Unionism and Reform Legislation, 1890–1911" (Master's thesis, Auburn University, 1979); Holman Head, "The Development of the Labor Movement in Alabama prior to 1900" (Master's thesis, University of Alabama, 1955); Nancy R. Elmore, "The Birmingham Coal Strike of 1908" (Master's thesis, University of Alabama, 1966); and Inez Norton Edwards, "Emmet O'Neal: Alabama Governor, 1911–1915" (Master's thesis, Auburn University, 1957).

Index

Index

Index

Index

Index

About the Authors

Robert David Ward is Professor Emeritus of History, Georgia Southern University, and William Warren Rogers is Professor of History, The Florida State University. Both received their B.S. and M.S. degrees in history from Auburn University and their Ph.D. degrees from The University of North Carolina at Chapel Hill.

They are co-authors of *Labor Revolt in Alabama: The Great Strike of 1894* and *August Reckoning: Jack Turner and Racism in Post–Civil War Alabama*. Other works by Rogers include *The One-Gallused Rebellion: Agrarianism in Alabama* and a three-volume history of Thomas County, Georgia.